SOCIAL BEHAVIOUR IN FLUCTUATING POPULATIONS

Studies in Behavioural Adaptation

Series Editor: John Lazarus, Department of Psychology,
             The University of Newcastle upon Tyne

Gulls and Plovers: The Ecology and Behaviour of Mixed-Species Feeding Groups
C.J. Barnard and D.B.A. Thompson
(Croom Helm, London & Sydney/Columbia University Press, New York)

Modelling in Behavioural Ecology: An Introductory Text
Dennis Lendrem
(Croom Helm, London & Sydney/Discorides Press, Portland, Oregon)

Primate Social Systems
Robin Dunbar
(Croom Helm, London & Sydney/Cornell University Press, New York)

SOCIAL BEHAVIOUR IN FLUCTUATING POPULATIONS

ANDREW COCKBURN

Department of Zoology,
Australian National University

CROOM HELM
London • New York • Sydney

© 1988 Andrew Cockburn
Croom Helm Ltd, Provident House, Burrell Row,
Beckenham, Kent, BR3 1AT

Croom Helm Australia, 44-50 Waterloo Road,
North Ryde, 2113, New South Wales

Published in the USA by
Croom Helm
in association with Methuen, Inc.
29 West 35th Street
New York, NY 10001

British Library Cataloguing in Publication Data

Cockburn, Andrew
   Social behaviour in fluctuating populations.
   — (Studies in behavioural adaption).
   1. Vertebrate populations 2. Social
   behaviour in animals
   I. Title II. Series
   596'.05248   QH352

   ISBN 0-7099-3426-2

Library of Congress Cataloging-in-Publication Data

Cockburn, Andrew, 1954–
  Social behaviour in fluctuating populations / Andrew Cockburn.
    p. cm. — (Studies in behavioural adaptation)
  Bibliography: p.
  Includes index.
  ISBN 0-7099-3426-2
  1. Social behavior in animals. 2. Animal populations.
3. Behavior evolution. 4. Animal behavior. I. Title. II. Series:
Studies in behavioral adaptation.
QL775.C573   1987
591.5'248 — dc 19   87-15747
ISBN 0-7099-3426-2

Printed and bound in Great Britain
by Billing & Sons Limited, Worcester.

Contents

Series Editor's Foreword		vii
Preface		ix
Chapter 1.	Introduction	1
Chapter 2.	Habitat Heterogeneity	28
Chapter 3.	Mating Options	52
Chapter 4.	Territoriality, Space Use and Mating Systems	73
Chapter 5.	Life History Evolution: Growth and Puberty	100
Chapter 6.	Life History Evolution: Reproductive Strategies	130
Chapter 7.	Dispersal	151
Chapter 8.	Social Behaviour and Population Regulation: a Synopsis	173
References		189
Index		236

*For Dennis Chitty. For agreeing to disagree
over whether we agree*

Series Editor's Foreword

In the early years of this century a Scottish doctor speculated on the evolutionary origin of human tears. It seemed to him that with the increase in brain size and cognitive powers of our early ancestors many events in the struggle for existence would be just too distressing to witness. How comforting then for a mother, distraught by the sight of her child being devoured by a lion, to cloud her vision with a flood of tears!

Just so, though if the good doctor had pondered further, the following picture might have occurred to him, comfortable in his speculative armchair, and given him some pause for thought.

These stories do not, of course, get us very far in explaining the evolution of tears – or anything else – but they do remind us how far the study of behavioural adaptation had come this century. This is, in fact, an exciting time for students of behaviour. The last twenty years has seen a great advance in the theoretical armoury for tackling problems of behavioural evolution and adaptation, and a parallel expansion in empirical studies, particularly in the field. The concepts of inclusive fitness and evolutionary stability, for example, have helped to explain major features of social behaviour and have generated entirely new questions for the field worker to examine. Cost-benefit analysis and optimization theory have done the same for behaviour in general, and links with population biology and population genetics are becoming stronger.

The heady days which saw the birth of behavioural ecology and sociobiology are now over, the new concepts have been refined and consolidated, and field data and comparative studies have accumulated at an impressive rate. Now seems a good time to take stock, to review the state of the art and to point some directions for future work. These are the aims of the present series, which will examine questions of behavioural adaptation and evolution in depth. As for our intended readership, we hope that all those interested in such problems, from advanced undergraduate to research worker and lecturer, will find these books of value. Some contributions to the series will encompass particular areas of study, reviewing theory and data and presenting fresh ideas. Others will report the findings of new empirical studies of an extensive nature, which make a significant contribution by examining a range of interrelated questions. The richness, but also the difficulty, of functional enquiry results from the multiple effects of selection pressures and the complex causal relationships between the behavioural responses to evolutionary forces. Studies which measure a comprehensive set of behavioural attributes, and their ecological correlates, are therefore particularly valuable.

Not all taxa have been illuminated equally by the new concepts of behavioural ecology. Birds, fish, insects and large diurnal mammals have received most attention, not only because of the richness of their social lives but also simply because they are relatively easy to observe in the wild. While small, secretive mammals such as voles and lemmings have yielded fewer of their secrets to the behavioural ecologist, population biologists have spent many thousands of man-hours trapping and studying them in an attempt to understand the cycles in population numbers that characterize many species. Perhaps as a result of this focus on the causes and consequences of population cycles, however, a thoroughgoing analysis of the evolution of social behaviour and life history traits of small mammals has yet to be made, in spite of the great amount of data available.

In the present book Andrew Cockburn argues that the emphasis on population density as a causal agent in the evolution of social behaviour in these fluctuating populations has directed attention away from other factors of proven influence in the more easily studied taxa with stable populations. He seeks to rectify this state of affairs by examining features of dispersion pattern, reproduction, parental investment and life history, and attempting to understand their evolution using theories developed within the field of behavioural ecology. As a result he provides a less parochial and more parsimonious view of the evolution of the social systems of small mammals, and of other vertebrate species which cycle in abundance. His analysis gives a much-needed spur to the work of both population biologists and behavioural ecologists on these questions and should help to bring about a closer integration between the two.

John Lazarus
July 1987

Preface

In this text I consider the evolution of social behaviour and parental investment in vertebrate populations which fluctuate dramatically in abundance. The book is prompted by a number of considerations. First, I freely admit to being susceptible to what Charles Elton (1942, p. 8) called '... a realm of pleasant personal feelings that come from the contemplation of those delicate small warm bodies, the free busy life under hedge, or crossing stems of grass, the subterranean energy and skill that beat us so often at our own games.' Second, I sense a collapse of the optimism that led Elton and Dennis Chitty to suggest that the study of populations of small mammals which undergo regular irruptions would elucidate the causes of regulation of all populations. The study of small mammals has been swept from a position of primacy in empirical ecology by the undoubted success of new subdisciplines such as behavioural ecology, the principles of which are difficult to apply to nocturnal secretive mammals. Third, despite the healthy attempts in recent years to overcome these difficulties and integrate data from small mammal populations with the perspectives of microevolutionary theory, many of these attempts at synthesis are marred by the blinkering and circular assumption that the most important selective agent acting on small mammal populations is the density of conspecifics. Because the literature on small mammal populations is so formidable, it has proved debilitatingly difficult to maintain contact with both a literature which was originally concerned with demography and a newer literature concerned with behavioural adaptation.

I attempt to overcome these difficulties with a selective review of patterns of life history and social behaviour in the microtine rodents, drawing where appropriate on the related literature on snowshoe hares and tetraonids (grouse and ptarmigan), which also cycle in abundance. The review seeks to identify those traits which are characteristic of fluctuating populations, and those which are characteristic of higher vertebrate populations generally. Having hopefully thus escaped the circularity inherent in a narrow focus on population fluctuation, I try to relate the patterns of variation to the growing theoretical literature on the evolution of life histories and social behaviour. Although generally avoiding the debate on the most important causes of population regulation, I conclude by examining the validity of the, now widely held, belief that changes in 'individual quality' may generate cyclic fluctuations in abundance. While the emphasis is theoretical rather than practical, the treatment is largely non-mathematical, reflecting both my own shortcomings and the desire to attract as broad an audience as possible. Relevant mathematical literature is referenced where appropriate, without repeating the formal development of models.

My analysis of the literature is unfortunately limited in two ways. First, most of the articles I use as primary sources have been published very recently. This does not imply any particular faith in modernity. Instead it represents a way of abstracting a vast literature. I do recommend that readers tackle the excellent summary by Krebs and Myers (1974) and Elton's (1942) entertaining monograph as complementary to this review. Second, I do not read Russian or Finnish. The task of translating primary sources would have delayed preparation of this manuscript by many months. The high esteem in which I hold the translated Russian and Finnish references to which I have access leads me to suspect that some ideas may have been developed long before the preparation of this text. If this is the case, I apologise to the original authors.

Most of the ideas expressed had their genesis during the years when I was a postdoctoral fellow at the Museum of Vertebrate Zoology at the University of California at Berkeley, the Department of Zoology at Monash University, the CSIRO Division of Wildlife and Rangelands Research and the Research School of Biological Sciences at the Australian National University. I am most grateful to Hugh Comins, Charley Krebs, Tony Lee and Bill Lidicker for their enthusiastic encouragement and measured scepticism during our collaboration. During the course of writing my work has been generously supported by the Australian Research Grants Scheme and the Australian National University Faculties' Research Fund.

My field knowledge of the beasts I describe has been restricted to one and a half years in the Berkeley hills, and visits to the field sites of other workers. For showing me the wonderful world of microtines, thanks to Bill Lidicker, Dennis Chitty, Heikki Hentonnen, Jussi Viitaala, Johan Tast and Henrik Wallgren. For critical discussion and help with the literature on the diverse topics described herein, I thank Charley Krebs, Bill Lidicker, Rick Ostfeld, Frank Pitelka, Erik Framstad, Nils Stenseth, Painey Pearson, Dennis Chitty, David Happold, Tony Lee, Steve Emlen and Bill Foley. Robert Moss and John Lazarus read the entire manuscript and made numerous invaluable suggestions. For access to preprints of useful references I thank Nick Barton, Joanna Gliwicz, Tony Lee,

Dale Madison, Ian McDonald, Bill McShea, Rick Ostfeld and Nils Stenseth. Milton Lewis and Rebecca Torrance drew most of the illustrations. John Lazarus and Tim Hardwick provided the incentive to attempt this exercise (a contract), and the encouragement to see it to completion (patience is a great stimulus for guilt).

Andrew Cockburn, Canberra

Chapter 1
Introduction

The dramatic fluctuations in the numbers of some mammals and birds in the northern latitudes of Eurasia and North America have proved disastrous for farmers, rich sources of folklore, frustrating for hunters and fascinating for biologists. In attempting to review the behaviour of these animals, it is worth saying at the outset that I do not intend to concern myself primarily with population regulation. The evidence that population regulation in microtine rodents, in some hares and in some ptarmigan and grouse has multiple causes is now overwhelming. However, research into the nature of population regulation has been one of the most substantial endeavours by ecologists, and remains central in the development of ecology. Like the science of ecology as a whole, the study of population regulation of vertebrates which fluctuate in abundance has progressed from an initially descriptive phase to a point where ideas are subjected to experimental scrutiny, and where mathematical logic has been used to generate more or less precise predictions of how the real world should behave. One theme underlies this research: if the local density of a population varies from year to year by several orders of magnitude, the social pressures faced by the animals in question must also vary dramatically. In turn, social behaviour may become an important selective influence, but a selective influence which varies in intensity. This may shape the behavioural and reproductive strategy of the animals themselves. These influences are the subject of my analysis.

Although early ecologists, to the point of formal definition, were concerned with the 'distribution and abundance' of plants and animals, a theoretical basis for ecology has remained elusive. Historians of science recognise two approaches to ecology:
(1) an Aristotelian idealism which perceives communities of organisms as

'individuals' with properties which distinguish them from the sum of their parts; and

(2) a Darwinian view, which ascribes the properties of organisms to their unique evolutionary histories, and as a consequence of natural selection for characters which favour the survival and reproduction of individuals.

New vigour has been added to the evolutionary perspective on ecology by the realisation that behaviour, like morphology and physiology, is eminently suited to Darwinian analysis. Although this point was hardly lost on Darwin himself, attempts to define an evolutionary basis for the study of social behaviour remained controversial until a series of important papers and monographs published in the 1960's and 1970's established that:

(1) altruism and cooperative behaviour could be understood in terms of the genetic benefit to individuals, instead of in terms of benefit to the group or species, as had widely been suggested (Hamilton 1964; Williams 1966a; Maynard Smith 1964);

(2) that because behaviour is often extremely variable within and between species, it may represent an unusually *good* opportunity to examine the operation of natural selection (Crook 1964, 1965; Lack 1968; Wilson 1971); and anticipated rather than treated as unfortunate noise in the system under investigation (Maynard Smith and Price 1973). These considerations led to the genesis of a subdiscipline known most widely as *behavioural ecology*. In a recent review, John Krebs (1985), one of the chief architects of this synthesis, provided an analysis of these developments, and suggested that it is now time that behavioural ecologists sought to synthesise their science with advances made by students of population regulation and life history evolution (see also Smith and Sibly 1985).

Microtine rodents, snowshoe hares, and grouse and ptarmigan offer an unusual opportunity to forge a synthesis of this sort. The data on population fluctuations is unusually rich and easily gathered, and the lives of the animals are rather short, facilitating life history analysis. Although much of the behaviour of small rodents remains hidden from scientific investigation, optimism that changes to intrinsic characters of populations, particularly those mediated through social interactions, underlie population fluctuations, has led to an increasingly sophisticated literature on the social behaviour of these species. Unfortunately these advantages have combined to limit the extent to which a synthesis of information has been attempted. The sheer enormity of the microtine literature inevitably leads to a narrowing of focus among researchers, as well as considerable tearing of hair and gnashing of teeth. The lessons to be learned by comparison with other animals are missed, particularly in peak years, when just trapping the beasts is a full time occupation. During the past ten years, with the exception of a brief sojourn confronting the animals themselves, I have studied the evolution of social behaviour and life histories of small mammals other than microtines. The analogies between microtines and these animals, all of which are comparatively rare Australian rodents and marsupials, prompt this review.

But first a disclaimer. I do not attempt, and could not hope to attempt, a complete review of the microtine literature. Excellent catalogues and reviews are available elsewhere, both of the literature as a whole (Krebs and Myers 1974; Tamarin 1985) and of specific topics (Stenseth 1977; Merritt 1984). Rather I focus on evolution of social behaviour and life history, and only enter the arena of population fluctuation where these three factors interact. But first it is necessary to identify the attributes which have led to the intense interest in microtine rodents by briefly reviewing what is known of their population dynamics.

Fluctuation and Cycling

The scale of population fluctuations in microtine rodents has to be witnessed to be comprehended. Getz (1985) mentions one outbreak of *Microtus montanus* in western North America where densities exceeded 25,000 voles/ha, and where in one county in Nevada over 10,000 ha of lucerne were completely destroyed. Yet at other times the animals are so scarce and patchily distributed that they might easily be classified as rare. The numbers of many small mammal and avian species fluctuate through time, and all zoogeographic regions have species which occasionally reach plague proportions (e.g. *Rattus* in Australia (Carstairs 1976); *Mastomys* in Africa (Davis 1964); *Oryzomys* in the Neotropics (Gilmore 1947); and the numerous Nearctic and Holartic microtine species). However, some microtine populations appear to exhibit periodical lows and highs more or less regularly, leading to the suggestion that these fluctuations are cyclical. Although Krebs and Myers (1974) concluded that cyclical fluctuations were characteristic of all microtines, abundant evidence now suggests that several species exhibit annual fluctuations associated with seasonal reproduction in some areas, occasional outbreaks or sustained high densities at other sites, and true multiannual cyclic fluctuations elsewhere. Unfortunately, rigorous statistical criteria to distinguish whether the periodicity of fluctuations in numbers of microtines and other small mammals are any more regular than in other species have only recently been applied to long-term data sequences (Garsd and Howard 1981; Finerty 1980; Hansson and Hentonnen 1985a; Hentonnen, McGuire and Hansson 1985; Taitt and Krebs 1985; Williams 1985). These preliminary analyses allow certain conclusions. First, it now appears that simple generalisations about the 'cyclicity' of a particular species of the type attempted by Krebs (1979) are probably of little use. Second, although resolution of the nature of this variation requires further study, it does seem possible to conclude that the regularity and amplitude of fluctuations within the microtines increase with latitude, as does the synchrony in fluctuations between species (Figure 1.1). Last, repeated samples within species suggest that there is continuous variation from strict multiannual cyclicity to more irregular fluctuations or annual population cycles (Figure 1.1). I do not believe it is useful to propose formal dichotomies as suggested by Taitt and Krebs (1985), who argued that cyclic populations and those which fluctuate annually or

Figure 1.1: Cyclicity in populations of Clethrionomys glareolus *from different latitudes in Fennoscandia. Data are from Hansson and Hentonnen (1985a). S, the index of cyclicity, is calculated according to the formula*

$$S = \sqrt{(\Sigma(\log N_i - \overline{\log N_i})^2/(n-1))}$$

where N_i is a quantitative density index at the same time each year (August-October data were preferred), and n is the number of years in the sample

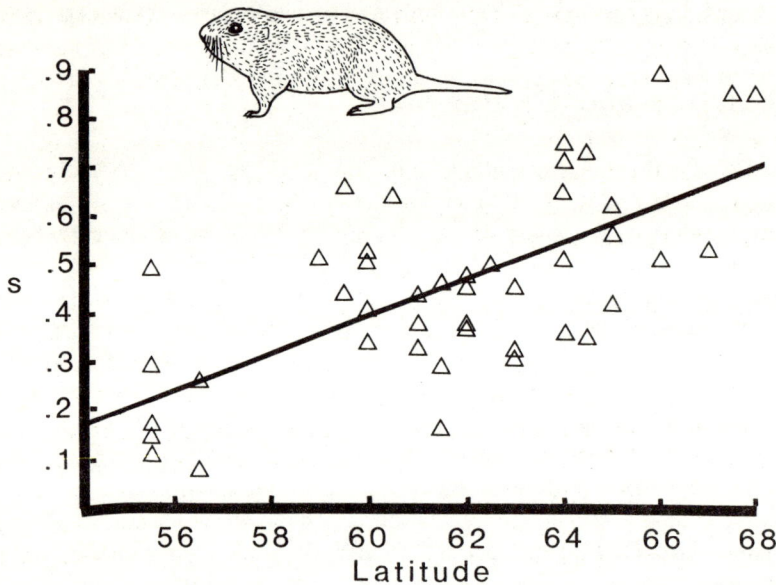

irregularly are fundamentally different.

I will, however, concentrate on those populations where true cyclicity is most pronounced, as I am inclined to agree with Chitty, Krebs and others that these circumstances offer a special opportunity for research because of the predictability and inevitability of both population increase and decline. It is possible, therefore, to discern three problems of population regulation. The first, stated most forcefully by Krebs and Myers (1974), is why populations do not increase in size indefinitely? This may be restated as concern with the amplitude of fluctuations. The second question is what factors cause regularity and synchrony in the fluctuations? This may be restated as concern with the periodicity of fluctuations. The final question, with which I will be principally concerned, is the extent to which violent fluctuations alter the selective milieu in which the animals live.

What Determines the Amplitude of Fluctuations?

A recent review by Taitt and Krebs (1985) elegantly summarises our present knowledge, and their conclusions (pp. 567-8) may be reproduced verbatim:

Microtus numbers increase when extra food is provided experimentally to field populations, but no one yet has prevented a cyclic decline by food addition. It is not yet certain whether plant secondary compounds play any role in vole cycles. Predation interacts with cover to affect vole numbers, and predators can take large numbers of voles under certain conditions. Predators may prolong the phase of low vole densities but it is not clear that they can generate cycles.

Spacing behavior operating through differential dispersal may be a key element in the adjustment of *Microtus* densities to available resources. Surplus voles exist in some populations, but we do not know what role such voles play in generating the population dynamics observed. Spacing behavior could be under both genotypic and phenotypic control, which suggests a multi-factor component in vole population dynamics. There is renewed interest in physiological responses of voles and lemmings to stress, and speculations about its effect on suppression of the immune-inflammatory system, especially at high population densities.

Periodicity

Hentonnen, McGuire and Hansson (1985) have recently compared the regularity of population fluctuations in Fennoscandia and shown that true cyclicity increases with latitude (Figure 1.1). Furthermore, sympatric microtines often irrupt and decline synchronously, as do their predators. These data are the strongest evidence that population regulation is driven by extrinsic rather than intrinsic factors (Hentonnen *et al.* 1985). However, the very large number of environmental circumstances which change with latitude in Fennoscandia obscure interpretation of the dominant causes of multiannual periodicity.

One recent line of evidence that does appear to implicate intrinsic factors in the periodicity of cycles is the observation that cycle length in a variety of homeothermic vertebrates is allometrically related to the body size of the species (Figure1.2; Peterson, Page and Dodge 1984; Calder 1983), subject to the caveat that cycle period of some species varies from site to site (e.g. Williams 1985), perhaps because the dominant members of any community may influence predator dynamics sufficiently to drag other species through population cycles. The function derived by Peterson *et al.*

$$\text{Cycle period (years)} = 8.15\, M^{0.26} \quad \ldots 1.1$$

where M is body mass in kilograms, has an exponent similar to the common exponent (0.25 or 0.75) derived for allometric relations between body mass and many physiological and life history parameters (Peters 1983; Calder 1984). This has led to speculation that the periodicity of cyclicity is simply dependent on the reproductive capacity of the species in question, and it is reasonable to anticipate some degree of cyclicity in most populations of vertebrates. Cyclicity may be most marked where a founding event or bottleneck through a period of very low density imposes some initial synchrony on patterns of natality and mortality. This suggestion is supported by the observation that a cycle of the predicted

Figure 1.2: Population cycle period as a function of body mass for 41 species of birds and mammals. After Peterson, Page and Dodge (1984)

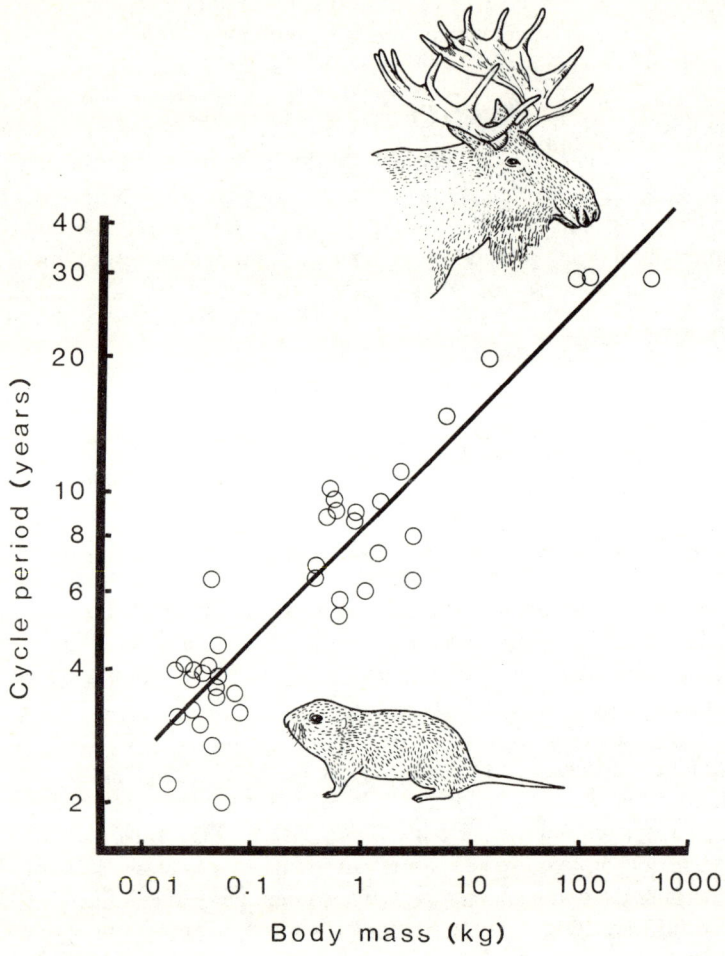

length occurred during recovery from near-extinction by one of the world's rarest birds, the whooping crane (*Grus americana*) (Boyce and Miller 1985). These intriguing data also suggest that high density is unnecessary for cyclicity. Indeed, it seems reasonable to conclude that the unusual characteristic of microtine and snowshoe hare cycles is their amplitude, and not the phenomenon of cyclicity itself. If these conclusions are true, the source of synchrony in northern Fennoscandia and the northern Nearctic might be the intensity of the bottlenecking during the adverse conditions experienced in the winter.

The Social Basis of Population Regulation

The notion that some populations regulate their numbers below their capacity for reproduction is well established (C.J. Krebs 1985a), but the relative importance of intrinsic and extrinsic influences on population numbers has not been adequately resolved. Indeed, resolution of this difficulty is increasingly being recognised as a consequence of semantic and conceptual confusion, rendering the historical debate on this topic unnnecessary. Krebs (1978) points out that self-regulation may be physiological or behavioural and may be genotypic or phenotypic, leading to four possible combinations of hypotheses about self-regulation (Table 1.1). In the following text, I illustrate data which test the validity of each of the major hypotheses implicating social behaviour in population regulation. While the data I describe are a fragmentary account of the literature as a whole, attempts to reject each of the major hypotheses serves to introduce the behavioural variation in fluctuating populations of vertebrates which will be considered throughout the remainder of this text.

The Chitty Polymorphic Behaviour Hypothesis

Several hypotheses implicate changes in the patterns of social behaviour as a primary determinant of population regulation. The most influential of these ideas was originally proposed by Dennis Chitty (1958, 1960, 1967) and suggests (1967, p. 51) that '... all species of animals have a form of behaviour that can prevent unlimited increase in population density', and (1967, p. 72) 'Mechanisms for the self-regulation of animal numbers are thought to be a consequence of selection, under conditions of mutual interference, in favour of genotypes that have a worse effect on their neighbours than *vice versa*.' The testability of the Chitty hypothesis and its corollary predictions has been subject to a valuable review by Krebs (1978) (Table 1.2). It is clear that the essential empirical elements

Table 1.1: Possible combinations of self-regulation mechanisms (after Krebs 1978) and the hypotheses of social regulation historically applied to fluctuating populations of vertebrates

	Physiological	Behavioural
Genotypic	—	Polymorphic behaviour hypothesis (Chitty 1967) Kin selection hypothesis (Charnov and Finerty 1980) Heterozygosity hypothesis (Smith, Garten and Ramsey 1975)
Phenotypic	Stress hypothesis (Christian 1971)	Behaviour hypothesis (numerous sources) Social fence hypothesis (Hestbeck 1982)

8 INTRODUCTION

Table 1.2: Predictions of the Chitty Polymorphic Behaviour Hypothesis and its derivatives (e.g. Charnov and Finerty 1980), and the degree to which they differ from other hypotheses of population regulation (modified from Krebs 1978). I have deleted predictions no longer believed to be true by the chief advocates of this hypothesis (Chitty 1967; Krebs 1978; C.J. Krebs, pers. comm.)

A. Predictions not unique to the hypothesis

(1) Spacing behaviours will be less common or intense in increasing populations than in declining populations

(2) The genetic composition of a population differs in increasing, stable and declining populations

(3) If animals are prevented from interacting adversely, they should go on increasing until they run out of food

(4) Animals reared in isolated cultures should resemble those in expanding, newly introduced or severely exploited populations, rather than those in stationary or declining populations

B. Unique to hypotheses of intrinsic or self-regulation

(1) Numbers should continue to increase if animals from an increasing population are transferred to an area from which a declining population has been removed

(2) Numbers should continue to decline if animals from a declining population are transferred to a new area from which other animals have been removed

C. Unique to the Chitty Polymorphic Behaviour Hypothesis

(1) Animals present in stationary or declining populations are selected for their ability to survive the effects of mutual interference. Animals in increasing populations are not under selection for interference, but are otherwise better adapted. When both animals co-occur, the former type should be most fit at high densities and the latter type most fit at low densities. Populations contain individuals reflecting two adaptive peaks, one containing genotypes adapted for reproduction and colonisation and the other containing genotypes adapted for conditions of mutual interference

(2) Spacing behaviours must be heritable, and their expression not swamped by environmental effects. Natural populations can be manipulated by selection experiments on individuals which reflect the two adaptive peaks

(3) Plague populations develop when the individuals of the high-density adaptive peak are lost and only increase genotypes remain in the population

---

of the theory are strong heritability for spacing behaviour and strong density-dependent natural selection for these behaviours. On a theoretical level it is necessary to demonstrate that selection of this sort will generate the fluctuations observed in nature. In the ensuing section I will argue that data from fluctuating populations of grouse and small mammals now permit falsification of both the Chitty hypothesis and its derivatives.

Mathematical plausibility.
There is now wide consensus that despite the restrictive assumptions involved in

mathematical analyses of biological problems, simple models are useful in exploring the feasibility of hypotheses which have a numerical basis. Mathematical analyses of population fluctuations have most commonly been directed towards determining conditions which permit regular fluctuations, or cycles (for a comprehensive analysis see Nisbet and Gurney 1982). Defining necessary or adequate conditions for cyclicity has not proved particularly difficult (May 1981a; Emlen 1984). Producing a biologically plausible model of the Chitty hypothesis has been a much more formidable task.

The principal postulate of the Chitty hypothesis concerns density-dependent selection on behaviour and its consequences for reproduction. This idea has been well treated mathematically, and Stenseth (1978a) points out that the notion of docile and aggressive phenotypes is analogous to the independently derived theory of r- and α-selection (MacArthur and Wilson 1967; Gill 1974). In elaborating this theory, Gill (1978a) identifies four responses to selection at high densities:

(1) escape from the crowded patch in space (dispersal);
(2) escape from the crowded patch in time (diapause);
(3) develop tolerance for the high density (K-selection);
(4) develop superiority through interference mechanisms (α-selection).

Unfortunately, theory does not yet contain useful *a priori* predictions of which outcome will be favoured by selection, leading to tautology in the interpretation of data pertaining to the theory (Stearns 1977; Chapter 7).

If we concede that mathematical theory does support the notion of α-selection in crowded environments, the next step is to investigate whether repeated selection for interference mechanisms at high density, and high reproductive rates at low density, can lead to the feedback necessary to produce cyclical population fluctuations. Because the Chitty hypothesis is principally concerned with genetic changes, the mathematical theory of population and quantitative genetics can be applied to this problem. Several models which analyse the Chitty hypothesis and assume a simple one locus, two allele control of aggressive behaviour have been proposed. Some of these models support the central premises of the Chitty hypothesis, but unfortunately either contain parameters with little biological meaning, or lack an explicit statement of the mathematics used to derive the conclusions. The first of these attempts (Dekker 1975) has been justifiably criticised by Nichols, Hestbeck and Conley (1979) and will not be discussed here. Hunt (1982) analysed dynamics of a population subjected to 'slow' density-dependent selection and concluded that cycling could occur given some restrictive conditions which appear more related to mathematical convenience than biological reality. One important conclusion was that animals homozygous for the docile allele must die out at high density for cyclicity. Page and Bergerud (1984) claim to have created a more biologically realistic model which adequately describes data they have collected for ten-year cycle of grouse. However, the mathematical basis of their model is not made clear (at least to me). The most comprehensive application of this approach to microtine cycles has been provided in a substantial series of papers by Stenseth and his co-workers. He concludes as the result of exhaustive analyses (1981, p. 27):

> I know of no other *realistic* models for Chitty's theory. The conclusion is therefore that *no satisfactory model has as yet been presented which shows that Chitty's presumed consequences are theoretically correct, when environmental variation is rejected as essential for generating the cyclic changes.*

Stenseth (1986) has more recently argued that the effect of spacing behaviour and aggressive interactions to which Chitty attributes importance will be to stabilise rather than destabilise population parameters.

Because there is no biological basis for one locus control of aggressive behaviour, Anderson (1975) correctly argued that the techniques of quantitative genetics are a more appropriate method of analysis. She provided a quantitative genetic analysis of the feasibility of the Chitty hypothesis and concluded that heritabilities for spacing behaviour and aggressiveness in the vicinity of 80 per cent were essential for cycling. With the single exception of a study by Singleton and Hay (1982) on house mice, there are no good data suggesting that such heritabilities are approached in nature. These shortcomings lend support to the view that the Chitty hypothesis is not a sufficient cause of cyclicity. However, it is worth pointing out that no modelling effort simultaneously incorporates both the notion of density dependence implicit in the Chitty hypothesis, and the concept of frequency dependence which is appropriate in studies of animal conflict (Maynard Smith 1982). The relevant theory is still in its infancy (Parker 1985), but may be worth pursuing.

Genetic evidence.
The most controversial empirical aspect of the Chitty Polymorphic Behaviour Hypothesis is the suggestion that there are two genetically distinct behavioural morphs in microtine populations which fluctuate in frequency according to the phase of the population cycle (Figure 1.3). I am aware of only two attempts to measure the heritability of spacing behaviour, dispersal and aggression in wild populations of small mammals. Anderson (1975) found no heritability for spacing behaviour in *Microtus townsendii*. The sire's behaviour could not predict any aspect of offspring behaviour, but there was some evidence of maternal effects in four of the nine behavioural variables she quantified. In an analogous study, De Poorter (1984) failed to find any evidence of heritability of spacing behaviour in snowshoe hares. Although quantitative genetic analyses are fraught with difficulty (e.g. Falconer 1981; Atchley 1984; Boag and van Noordwijk 1986), this line of research offers to improve considerably our understanding of the evolutionary dynamics of populations and constraints on the potential for natural selection (Lande 1982). However, preliminary data can not be construed to support the Chitty hypothesis.

More comprehensive data are available for populations of red grouse (*Lagopus*

Figure 1.3: The Chitty hypothesis as interpreted by Krebs (1978, 1985b). Note that the aggressiveness should peak at the end of the decline period

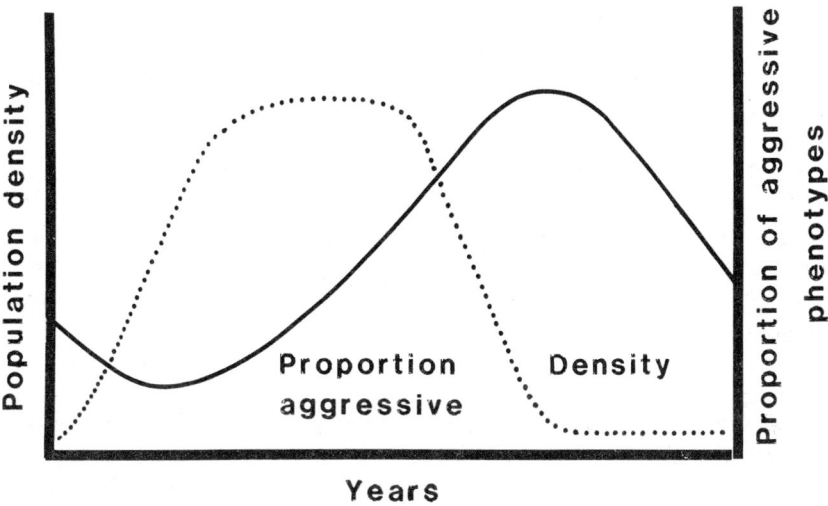

lagopus scoticus) which fluctuate in abundance with a period of six to eight years. Quantitative genetic studies of precocial bird populations are more tractable than in small mammal populations because genealogy is more easily determined, and early collection of eggs permits isolation of parents from their chicks before non-heritable maternal and/or paternal behavioural effects can develop. Estimates of heritability were available for a variety of life history and behavioural parameters of grouse (Moss and Watson 1982; Moss *et al.* 1982). Moss, Watson and Rothery (1984) took samples of eggs each year through an entire fluctuation and the chicks hatched from these eggs were reared in groups of 20 to 25 in standard conditions. In the following year, the chicks were bred in an aviary and their offspring raised in the same standard conditions.

Changes in population density and breeding performance on the study area are shown in Figure 1.4. Hatch weight of chicks, early survival and longevity as adults were not related to population density. Although heritable, changes with density in egg size and body weight were not due to genetic changes. There were inherent changes in dominance (Moss, Rothery and Trenholm 1985), and animals became more aggressive as the decline in numbers proceeded from 1974 until 1977 (Figure 1.5). Although aggressiveness and dominance are often only weakly related (Moss *et al.* 1979), in this study changes in aggressiveness did parallel changes in dominance level. If selection occurs, it appears to be during the decline phase rather than at the peak of numbers. This persistent increase in aggressiveness is clearly contrary to a central prediction of the Chitty hypothesis (Prediction C1; Table 1.2), though these data constitute the best (and possibly only) evidence for selection for different levels of aggressiveness in response to

*Figure 1.4: Breeding density (circles) and breeding performance, chicks reared per hen (triangles), in red grouse (*Lagopus l. scoticus*). After Moss, Watson and Rothery (1984)*

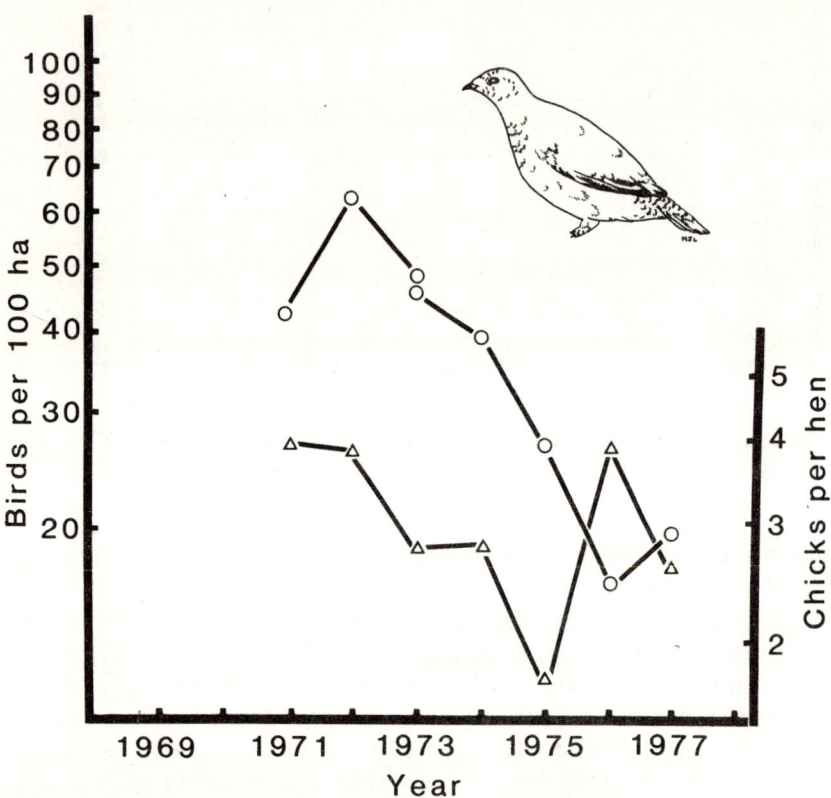

increased or decreased population density.

The other type of genetic data which have been applied to our understanding of microtine social behaviour are changes in allele frequencies detected by electrophoresis. The significance of fluctuations in the frequency of electromorphs has been the subject of a prolonged debate which is outside the scope of this review (see Thorpe 1982; Kimura 1983; Lewontin 1985; Mueller, Barr and Ayala 1985 for recent reviews). When biochemical geneticists began to use electrophoresis to identify the extent of genetic variation in populations, the frequency with which different forms of enzymes and proteins occurred within individuals and populations proved, if anything, to be an embarrassment to biologists attempting selectionist explanations of the variability (Lewontin 1974; Kimura 1983). This led to the formulation of the neutral theory of molecular evolution which proposed that many mutations are selectively neutral and that their frequency and probability of fixation in a given population is determined principally by random drift.

Figure 1.5: Aggressiveness scores of cocks from red grouse eggs collected from the population studied by Moss, Watson and Rothery (1984). The unit is the mean score for each clutch. Aggressiveness increased in parallel with dominance as the population declined. The vertical lines denote the standard error

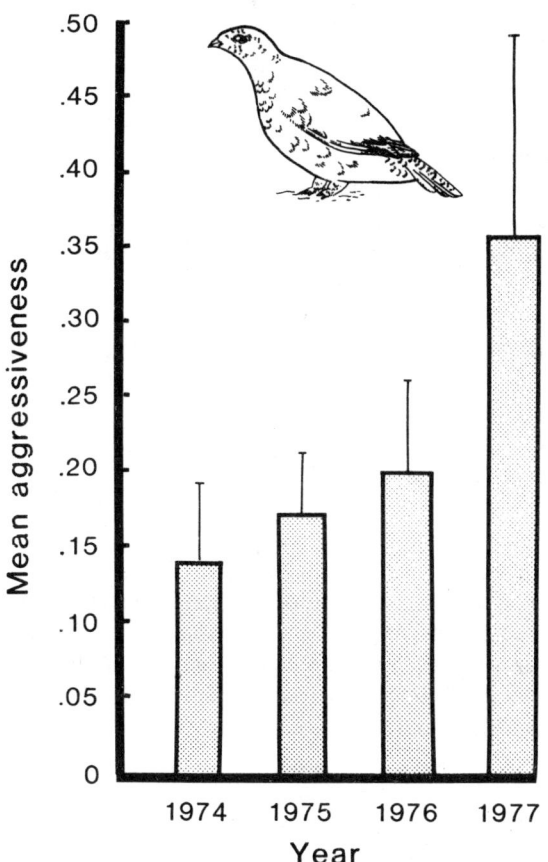

Attempts to show that genetic changes are indicative of selection pressure in small mammal population remain as controversial. First, it has been suggested that physiological condition may be more important than genotype as a cause of some of the variation in electromorphs of proteins commonly used in genetic surveys (Gulyaeva and Olenev 1979; McGovern and Tracy 1981, 1985). Transferrins and peptidases, which are often highly variable in mammal populations, were regarded as particularly suspect. Attempts to replicate these observations have met with little or no success (Mihok and Ewing 1983; Gaines and Gorman 1985; Lidicker, pers. comm.), but clarification of this issue should be a research priority. Although selection is often assumed to be present when allozyme frequencies change with population density, the roles of drift and selection have rarely been contrasted. Gaines and Whittam (1980) used a test devised by Lewontin and Krakauer (1973) to conclude that genetic drift was likely to have been more important than selection in determining allele frequencies at five loci in fluctuating populations of *Microtus ochrogaster*. The loci considered included transferrin and leucine aminopeptidase. Although no other data have been subjected to the criteria of the Lewontin-Krakauer test, the fluctuations in allele frequency observed by Gaines and Whittam did not differ greatly from those described in other microtine populations. In fact, Gaines, McClenaghan and Rose (1978) had originally interpreted the data to support the existence of selection on the loci under consideration.

It has recently been demonstrated that population fluctuations can reduce effective population size quite sharply (Motro and Thomson 1982). This result lends support to the notion that drift, which acts most strongly in small populations, is of considerable importance (see also Bowen 1982). A more intriguing possibility was raised by Gyllensten (1985), who suggested that at least some of the temporal heterogeneity in allele frequencies in a study of a fluctuating population of grouse (*L. lagopus*) was caused by differences in the propensity to sample family groups during particular phases of the cycle. The mechanism underlying this suggestion is not difficult to discern, particularly wherever the social system, differential survival of families and dispersal tendency are related to population density, as has been frequently implied for fluctuating populations.

It is also uncertain whether correlations between allele frequency and population density are indicative of cause or effect (Charlesworth and Giesel 1972; Charlesworth 1980). Gaines (1985) points out there is an extremely low probability of randomly identifying any loci causally related to cycling, yet positive correlations are frequently reported. This suggests that population fluctuations are facilitating changes in allele frequency, perhaps through drift, rather than a few loci driving population dynamics.

Some of the best evidence for selection sustaining differences in the frequencies of alternative electromorphs of particular allozymes comes from repeatable latitudinal trends. For example, in *Drosophila*, the genus of animals which has the best understood population genetics, there are trends in the frequency of allozymes of both esterase-6 and alcohol dehydrogenase which are repeatable both north and south of the equator, and between species (Gibson and Oakeshott 1980; Oakeshott, Chambers, Gibson and Willcocks 1981; Oakeshott *et al.* 1982). Such trends are likely to be the result of convergent evolution driven by selection. This interpretation is supported by increased understanding of the function of these enzymes. Alcohol dehydrogenase is important in diet, and esterase-6 secretion by males appears to suppress the libido and accelerate egg-laying of females mated by the male. It is not difficult to imagine latitudinal variation in the selective milieu acting on both these parameters.

Recent evidence suggests that activity of an esterase found only in cricetid rodents, the group to which microtines belong, shows latitudinal variation in *Clethrionomys glareolus* (Hall and Semeneoff 1985). This species shows latitudinal variation in density fluctuations (Figure 1.1). One form of the esterase (with high mobility) is more prevalent in the north, where fluctuations are most pronounced. This form was also most prevalent at high population density in a southern population (Wicken Fen, near Cambridge). Additional studies which replicate these results in different geographical localities (e.g. Fennoscandia), using different species and different allozymes, should be illuminating. There is now some evidence that in Sweden, *Microtus agrestis* taken from northern cycling populations and southern non-cycling populations differ in some aspects of behaviour (Rasmuson, Rasmuson and Nygren 1977). Although these results are promising, their connection to the Chitty hypothesis is uncertain, as they reflect interpopulation differences rather than intrapopulation differences.

Again, I would contend it is safe to conclude that available genetic data lend no support to the original formulation of the Chitty hypothesis, and that the first steps towards quantitative genetic analyses constitute a partial falsification, though the strength of this falsification is weakened by methodological difficulties associated with field analyses of heritability (Falconer 1981; Boag and van Noordwijk 1986).

Behavioural evidence.
Quantification of aggressive behaviour of small mammals in the field remains poorly developed, though the incidence of aggressive behaviour can be inferred from the incidence of wounding, particularly to the tail and flank glands. The most frequent direct measurements involve testing pairs of animals in neutral arenas or unfamiliar areas (e.g. Mihok 1981; Hofmann, Getz and Klatt 1982). There are compelling theoretical reasons and growing empirical evidence that suggest that these measures are all but meaningless, except where two animals displaced from their home range meet on unfamiliar ground. There is abundant evidence that pairwise conflicts in nature are generally settled without extreme escalation of the conflict to the point where one of the pair might be seriously injured. This follows the acquiescence of one of the pair. This is because rules which permit one individual to win a conflict according to some arbitrary convention (e.g. territory owners always win), are evolutionarily stable, even if there is no difference in ability between the two fighters (where there is an uncorrelated asymmetry). Evolutionarily Stable Strategy Theory calls this arbitrary convention 'Bourgeois' (Maynard Smith 1982).

Preliminary evidence that such a strategy may exist in microtine rodents has been provided by Bondrup-Nielsen (pers. comm.), who showed that conflicts in *Clethrionomys glareolus* are settled on the basis of territorial ownership. Gleason, Michael and Christian (1980) showed that changes in the aggressiveness of *Peromyscus leucopus* during the reproductive cycle were dependent on the encounter site. Animals in their home cage showed increases in aggressiveness towards a diestrous conspecific female throughout pregnancy and lactation when tested in their home cage, but in a neutral arena increased aggressiveness was only noticed early in lactation. In addition to these arbitrary uncorrelated asymmetries in pairwise contests, two types of predictable differences between contestants (correlated asymmetries) also mean that contests over territory may often be settled without escalation of conflict. The first of these is based on careful assessment of obvious differences in fighting ability. Such assessment ensures that a weaker animal will rarely attempt to tackle a stronger animal, which may in turn be more likely to hold a territory (for an example, see Clutton-Brock, Guinness and Albon's (1982) analysis of fighting in red deer stags). Last, owners may have a higher motivation to defend their territories because they have invested more time in finding appropriate places to nest and feed within it. In an intriguing study of great tits, Krebs (1982) was able to show that when a territorial bird was removed it was usually rapidly replaced by a non-territorial 'floater' from the population. If the original owner was released back on to the territory within a few days, it would easily displace the intruder. However, if the intruder was allowed to establish itself for a longer period of time it could effectively repel the original owner. For a brief period intermediate between these extremes, the motivation to fight for the territory was apparently similar, and escalated conflicts took place (Figure 1.6).

Because the results of encounter experiments are entirely dependent on the context in which they are conducted, the interpretation of other indices of behaviour measured in unusual contexts should also be treated with considerable caution. Despite these difficulties, the data which are available do not support the predictions of the Chitty hypothesis. Mihok (1979, 1981) claimed that 'docile' phenotypes predominated in a declining population of *Clethrionomys gapperi*, and animals were most aggressive during a period of population increase. However, his data can probably be only weakly related to the phase of the cycle, as the fluctuations at his site were dramatic, and the true 'increase' and 'decline' were probably not sampled. Mihok was forced to rely on observations of overwintered females, as males showed no changes in behaviour. Data from *Microtus ochrogaster* and *M. pennsylvanicus* are equivocal. Krebs (1970) suggested that male aggression was high in the peak phase but not in the decline phase, but Hofmann, Getz and Klatt (1982) detected no changes with density. C.J. Krebs (1985b) suggested that these differences were caused by the failure of Hofmann *et al.* (1982) to distinguish annual and cyclical changes in density. Although I agree that their data may not be an explicit test of the Chitty hypothesis, I disagree with Krebs that the distinction between cyclical and annual change is as clear as he implies.

C.J. Krebs (1985b) studied laboratory behaviour in breeding *Microtus townsendii* through various population conditions by taking individuals from field populations for two days and then returning them to their home range. Aggressiveness was positively related to population growth rate, in direct contrast to the predictions of the Chitty hypothesis. C.J. Krebs (1985b) concluded that, in general, docile microtine phenotypes predominate in the decline phase of populations. De Poorter (1984) obtained similar results for snowshoe hares which fluctuate over a period of about ten years in North America. However, aggression in

INTRODUCTION 17

*Figure 1.6: Prediction of a peak of escalation between residents and replacements at intermediate replacement times (after Krebs 1982; Parker and Rubenstein 1981). Krebs (1982) data from the great tit (*Parus major*) support this model*

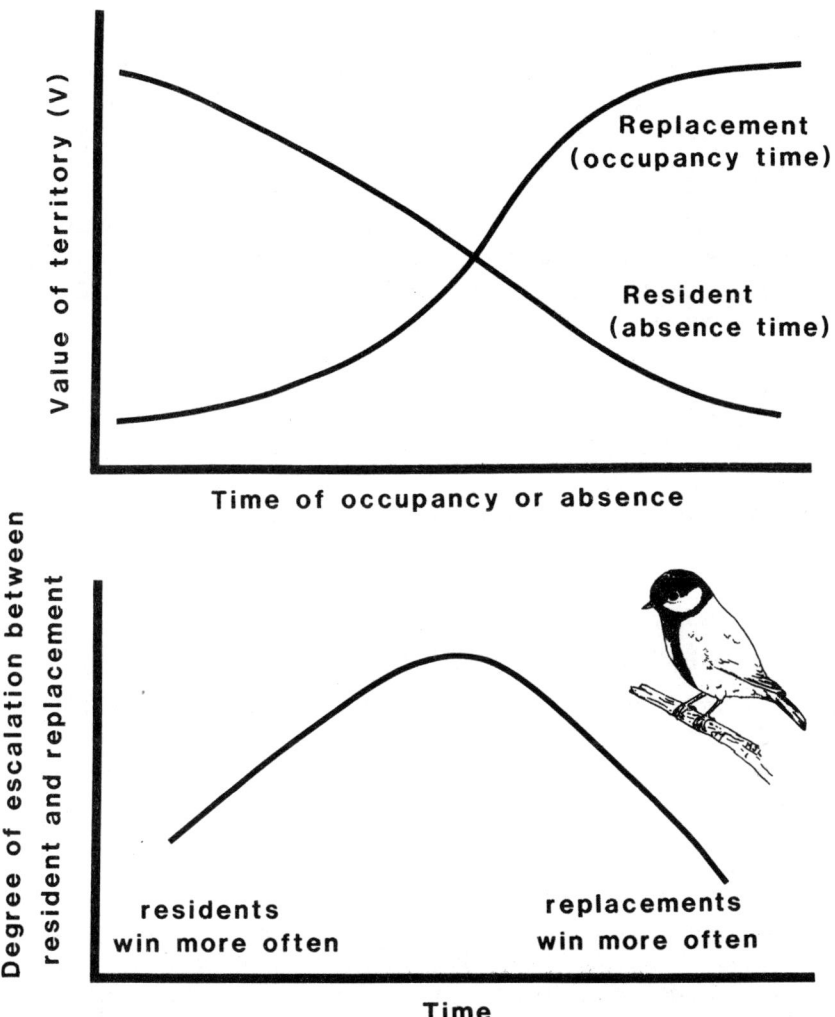

grouse increased well into the decline phase, weakening the generality of these conclusions (Figure 1.5). It is clear that neither pattern represents support for Krebs' (1978) interpretation of the predictions of the Chitty hypothesis.

Failure to predict features of cycling populations.
The other weakness of the Chitty hypothesis is the increasing evidence that the social systems of different microtine species are characterised by diversity rather

than similarity. Although this diversity forms the basis for several subsequent chapters, it is worth mentioning here that within the genera *Microtus* and *Clethrionomys*, it is possible to recognise the presence and absence of male territoriality, female territoriality and joint defence of territories by members of both sexes (Ostfeld 1985a). As a corollary, mating systems appear to include monogamy, male dominance and resource defence polygyny, and various degrees of promiscuity. By contrast with sharp interspecific differences in spacing behaviour, intraspecific differences in population fluctuations appear to be as great as interspecific differences. Further, there is compelling evidence that suggests that this pattern of territoriality exerts a stronger influence on levels of aggression than does population dynamics (e.g. Ostfeld 1985b; de Jonge 1983; Mihok 1981).

In summary, I would conclude that the difficulty in constructing biologically plausible models of the hypothesis, the absence of evidence for heritability of spacing behaviour, and contradictory but unpredicted behavioural patterns warrant abandonment of the Chitty hypothesis. However, the idea must be seen as a major step towards synthesising ecological and evolutionary theory, at a time when the major thrust of ecological research was directed to other, and I suspect less profitable, goals (see McIntosh 1980, 1985 for a review of evolutionary and non-evolutionary approaches to ecology). The results I have described do not constitute a rejection of the most important theoretical advance provided by Chitty, the notion that rapid environmental fluctuations may cause rapid but impermanent natural selection which may profoundly affect the properties of natural systems.

The Kin Selection Hypothesis

A derivative of the Chitty hypothesis is the suggestion by Charnov and Finerty (1980) and Finerty (1980) that variation in aggression is determined by the degree of relatedness between interacting individuals. Hamilton (1964) showed that an allele for altruism (behaviour leading to a cost in fitness to the actor and a benefit to the recipient) can only spread when:

$$rB-C > 0 \qquad \ldots 1.2$$

where r is the coefficient of relatedness between the recipient and the actor, C is the cost in fitness to the actor, and B is the benefit to the recipient as a result of the action. When *all other factors are equal* animals are predicted to direct cooperative behaviour preferentially towards relatives and aggressive behaviour towards strangers. Charnov and Finerty (1980) hypothesised that because at the start of a population cycle populations are likely to comprise small demes of related individuals, amicable behaviour should be prevalent. As populations expand, dispersal and fusion of demes would tend to decrease relatedness, and the increase in aggression would precipitate a population decline. Population subdivision at the following nadir would reinitiate this process.

Although it has been suggested that this hypothesis is more biologically and mathematically tractable than the Chitty model (Stenseth 1983; Warkowska-Dratnal and Stenseth 1985), I am unconvinced it warrants much further investigation. First, the Chitty hypothesis makes a number of predictions about behaviour in fluctuating populations which have been difficult to support empirically. The Kin Selection hypothesis appears to take these behavioural *predictions* as proven, and justify them theoretically. The circularity inherent in this attempt to bridge the gap between theory and the real world should be clear. Certainly, it takes considerable ingenuity to construe the Kin Selection hypothesis to support the observation that docile phenotypes predominate in declining populations of microtines.

Second, the ability to direct behaviour differentially according to relatedness is absolutely dependent on the capacity to recognise kin, and on the capacity to generalise recognition to distant relatives. While preliminary data support the notion that microtines can recognise their close relatives, at least part of the cues for recognition appear to be based on familiarity (Boyd and Blaustein 1985). Generalisation has not been studied, but there is no evidence of a decline in the duration of association between relatives at high density. Indeed, in some species family aggregations are more likely at high density (Jannett 1978, 1980).

The Heterozygosity Hypothesis

An hypothesis of identical form was suggested by Smith, Garten and Ramsey (1975). They argued that at low density populations are likely to consist of inbred demes, with homozygosity high within demes, but considerable genetic differentiation occurring between demes. As population density increases, genetic differentiation should be reduced as a consequence of dispersal and fusion of demes (the Wahlund effect). Heterozygosity should be maximal at and shortly after peak densities. Garten (1976) noted that aggression was correlated with heterozygosity in populations of *Peromyscus polionotus*, possibly because of the well documented phenomenon of hybrid vigour, or heterosis. According to this view, the increase in aggression would be an artefact of the breakdown of population subdivision, but the increase in spacing behaviours could initiate a decline. Electrophoretic evidence relating to this proposition is equivocal. Supportive data are available for two cycles in *Microtus californicus* (Bowen 1982; Lidicker 1985; Figure 1.7), and *M. agrestis* (Nygren 1980b), but Gaines, McClenaghan and Rose (1978) observed a deficiency of heterozygotes in *M. ochrogaster* as density increased.

Because the data required for the Chitty hypothesis are the same as those required to support the Heterozygosity hypothesis, I am dubious that these ideas are worth pursuing as a basis for fluctuations in social behaviour through the course of population fluctuations. However, the notion that changes in the quality of individuals is influenced by population structure and not selection warrants further investigation, and may indicate an underlying noise which should be taken into account in the design of experiments.

The Stress Hypothesis

A radically different set of ideas concerning the effects of individual differences on

Figure 1.7: The fixation index (F_{ST}), or reduction in heterozygosity of a subpopulation due to random genetic drift, averaged over four loci and four fluctuating subpopulations of Microtus californicus. *From Bowen (1982)*

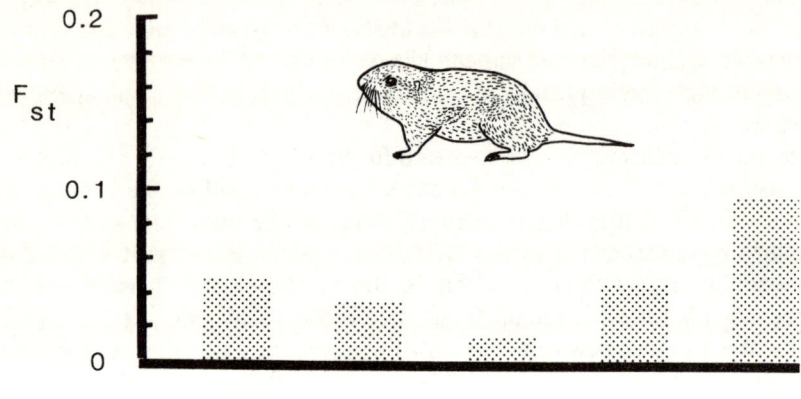

population processes has been formulated most clearly by John Christian in a number of exhaustive reviews (1961, 1971, 1975, 1978, 1980). Christian (1950) argues that increases in population density increase the level of agonistic behaviour experienced by individuals. Animals are predicted to respond to increased agonism with stress, and changes associated with stress in the neuroendocrine axis will impair survival, reproduction, and the reproductive capacity of their offspring. Such reproductive impairment might last through several generations. This hypothesis is similar to the Chitty hypothesis in its emphasis on changes in reproductive 'quality' of individuals lasting more than a single generation, but differs by implying that changes in quality are phenotypic rather than genotypic.

My discussion of this hypothesis is greatly influenced by the recent review by my friends and colleagues, Tony Lee and Ian McDonald (1985). It is first worth emphasising that animal physiologists attribute a different meaning to stress than do both plant physiologists (Harper 1982), and most other zoologists (Pickering 1961). In this text I retain the sense originally proposed by Selye (1936, 1946), who described stress as the set of responses exhibited by mammals in response to a variety of harmful environmental changes (stressors). Selye (1946) argued that stress contained both a specific response to the environmental agent, and a non-specific response, common to a wide variety of potentially adverse environmental factors, unrelated to their nature and determined only by their intensity. This general response, which he called the General Adaptation Syndrome, has three stages: (1) a stage of alarm; (2) a stage of adaptation; and (3) a stage of exhaustion.

The stage of alarm functions to mobilise energy reserves to counter the stressor. This includes cardiovascular effects and mobilisation of carbohydrate and fat in response to stimulation of the sympathetic division of the autonomic nervous system. In addition, secretion of glucocorticoids from the adrenal cortex is increased, both enhancing the effects of sympathetic nervous action, and causing mobilisation of tissue nitrogen through gluconeogenesis. The response is mediated by adenohypophyseal secretion of adrenocorticotrophic hormone (ACTH) which also causes mobilisation of fats in most mammals. The duration of this stage depends on the intensity and duration of the stressor, and the previous condition of the individual.

The stage of adaptation involves homeostasis in the presence of the stressor, and relies on other endocrine mechanisms which change energy balance from catabolic to anabolic, with a disappearance of much of the initial alarm reaction. The increased glucocorticoid secretion persists at a lower level, but the signs of increased sympathetic nervous activity disappear. Homeostasis is apparently maintained through an increase in energy turnover.

The final stage (exhaustion) occurs when the pituitary-adrenal response fails, possibly as a consequence of excessive ACTH action as a result of prolonged action of the stressor. The animal would then succumb either to the stressor, or as a direct result of adrenocortical collapse.

Lee and McDonald (1985) show that Selye interpreted the first two phases as adaptive responses to the stressor, and the third stage as a failure of the response. However, they point out that persistent high levels of adrenal glucocorticoid secretion have the additional deleterious consequences of suppression of the immune and inflammatory systems, increasing the susceptibility of the organism to a range of diseases and parasites which would ordinarily be controlled. There may also be adverse consequences of prolonged high adrenocortical activity on other aspects of homeostasis and on reproduction. Selye calls these unfavourable responses 'diseases of adaptation'.

The evidence for these effects has been reviewed by, among others, Christian (1980), Seabloom (1985) and Lee and McDonald (1985). First, the evidence for adrenal exhaustion in populations of mammals which fluctuate in abundance is very weak (e.g. Chitty 1959; see also Krebs and Myers 1974). This led Christian to change his emphasis from exhaustion to overactivity of the adrenals, particularly through the effects this would have on susceptibility to disease and parasites. The literature pertaining to this effect is weakened by the lack of a reliable means of determining adrenocortical activity. Variation in adrenal weights are sometimes sex-and species-specific and are not necessarily correlated with corticosteroid secretion rate or plasma corticosteroid concentration. The different zones of the adrenal cortex may be a more useful measure, but insufficient data are available to allow comment on species and sex differences in the structure of the adrenals. In addition, any measure which requires killing of study animals is likely to be unsuitable for attempts to correlate population structure with endocrine activity. Lee and McDonald (1985) consider a number of indices and recommend measurement of both the maximum corticosteroid binding capacity (MCBC) of high affinity corticosteroid proteins, and total corticosteroid concentration. The unbound or free corticosteroid concentration is the best available measure of susceptibility to stress. This claim can be illustrated with observation of stress-induced mortality in the small marsupial, *Antechinus stuartii* (Figure 1.8). Males show a sharp androgen dependent fall in MCBC during mating, so that free plasma glucocorticoid increases sharply, even though total corticosteroid levels increase only slightly (Figure 1.9). It is intriguing that high glucocorticoid levels usually reduce ACTH secretion before pathological effects become pronounced, but in *Antechinus* this feedback is impaired (McDonald, Lee, Than and Martin 1986).

The ultimate consequence of these elevated free corticosteroid levels is complete mortality of all males within two weeks of mating, providing the best evidence of a significant role for stress-related mortality in field populations (Lee and Cockburn 1985a). This mortality has been naively interpreted as altruism on the part of males by some authors (Diamond 1982). However, Lee and Cockburn (1985a, 1985b) have suggested that the persistent high plasma glucocorticoids during the mating period is beneficial to the males because the gluconeogenic mobilisation of tissue nitrogen as a source of protein reduces the time males need to feed during the frenetic search for mates during the period of mating (see also Scott and Tan 1985). There are close analogies between this phenomenon and the utilisation of tissue as an energy reserve through glucocorticoid-induced nitrogen mobilisation and gluconeogenesis in spawning Pacific salmon (Robertson and

Figure 1.8: Patterns of endocrine change influencing susceptibility to stress and gluconeogenic mobilisation of protein in Antechinus *spp. After Lee and Cockburn (1985a), McDonald et al. (1986)*

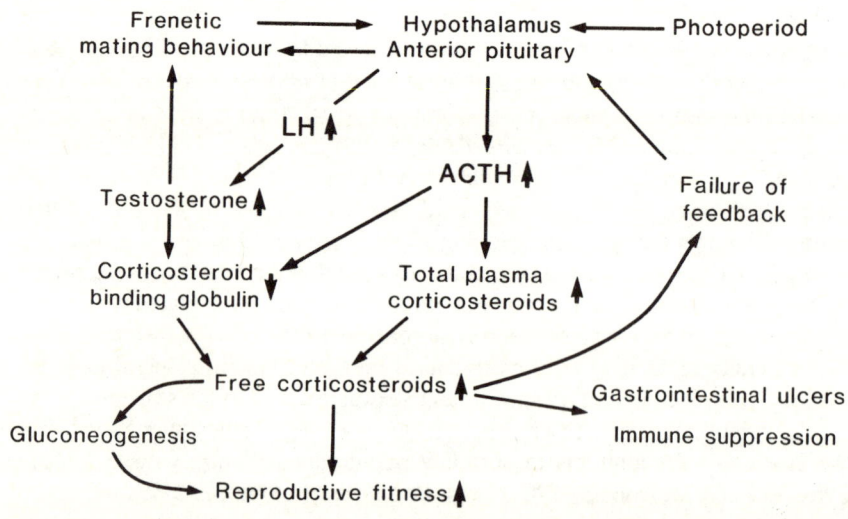

INTRODUCTION 23

Figure 1.9: Corticosteroid binding globulin and plasma corticosteroids in Antechinus, *males (above) and females (below). The mating season is denoted by the dark bar on the horizontal axis. Note that plasma corticosteroid levels in males are lower than those in females, but binding capacity in males in the mating season does not increase, or declines, in response to the increase in total plasma corticosteroids. After Lee and Cockburn (1985a)*

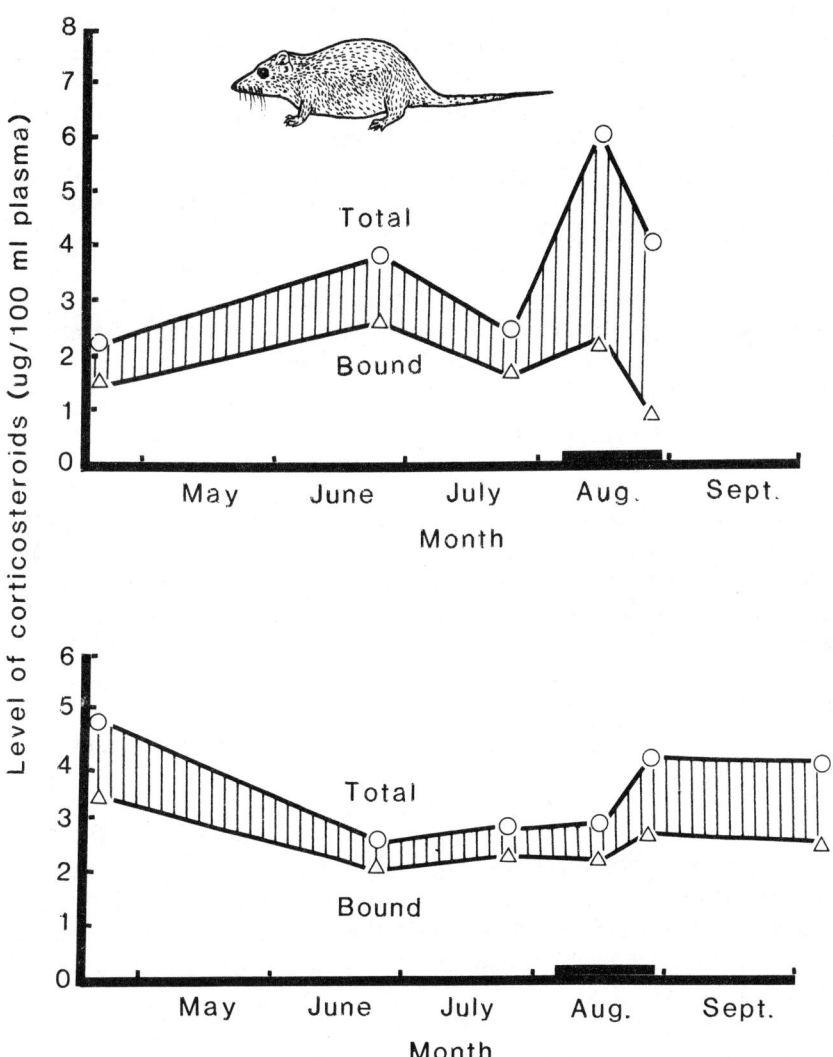

Wexler 1957; 1960; Hare and Robertson 1959; Robertson 1961).

If, as I believe, this interpretation is correct, male mortality in *Antechinus* cannot be construed as evidence for the Stress Hypothesis, in contrast to the original claim of Lee, Bradley and Braithwaite (1977). Rather than viewing high levels of stress as undesirable, it suggests that stress may be used as an adaptive response under conditions of intense competition for mates, particularly when food is in short supply, even if the cost of the stress is suppression of the immune and inflammatory systems. Lee and Cockburn (1985b) have suggested that these conditions are most likely in rodent populations during the synchronous onset of breeding at the start of the spring. Several authors have documented 'spring declines' in populations of small mammals (Fairbairn 1977; Krebs, Halpin and Smith 1977; Gurnell 1978; Krebs and Boonstra 1978; Cockburn 1981; Taitt and Krebs 1983), but their cause remains uncertain. Lee and Cockburn (1985b) have argued that the role of adaptive stress response in males during these declines warrants further investigation, but requires use of the correct indices of stress (see also McDonald and Taitt 1982).

Despite these interesting results, I agree with recent reviewers who conclude that there is little *field* evidence for the Stress Hypothesis as it was originally formulated by Christian (1970, 1978, 1980).

Social Fence Hypothesis

Hestbeck (1982) has recently devised a social model of population regulation which has its antecedents in multifactorial models proposed by Lidicker (1973, 1978). He argues that when animals are confined to small core areas, spacing behaviour regulates the population through emigration into unoccupied patches. As all available patches are saturated, the ability of emigration to regulate populations in the core areas is reduced. There is a change in the relative importance of aggressive interactions between members of small local groups and interaction between members of nearby groups. Ultimately the core area is 'fenced' by social pressure, as no further dispersal is tolerated. Under these circumstances, resource exhaustion occurs, and the population declines rapidly (Figure 1.10). There is experimental evidence to support at least some of the premises upon which this model is based. When physical barriers are constructed which frustrate emigration, microtine populations attain extremely high densities, overgraze the habitat and decline to very low levels or extinction (Krebs, Keller and Tamarin 1969; Boonstra and Krebs 1977). The fence effect can be eliminated by allowing dispersal through gateways (Gaines, Vivas and Baker 1979; Beacham 1980a, 1981a), or into areas of unsuitable habitat. For example, Tamarin, Reich and Moyer (1984) prevented immigration into an enclosure by fencing an area of grassland inhabited by *Microtus pennsylvanicus* but permitted emigration by allowing movement into a woodland area unsuitable for occupancy by voles. Although some studies have failed to document the fence effect (Verner and Getz 1985; Ford and Pitelka 1984), the methods used are open to question (Krebs 1986). In addition, there is some evidence that dispersal does decline in impor-

Figure 1.10: The social fence hypothesis of Hestbeck (1982). The importance for population regulation of intrinsic factors (spacing and dispersal) should decline, and the importance of extrinsic characters (food supply) should increase as the density increases

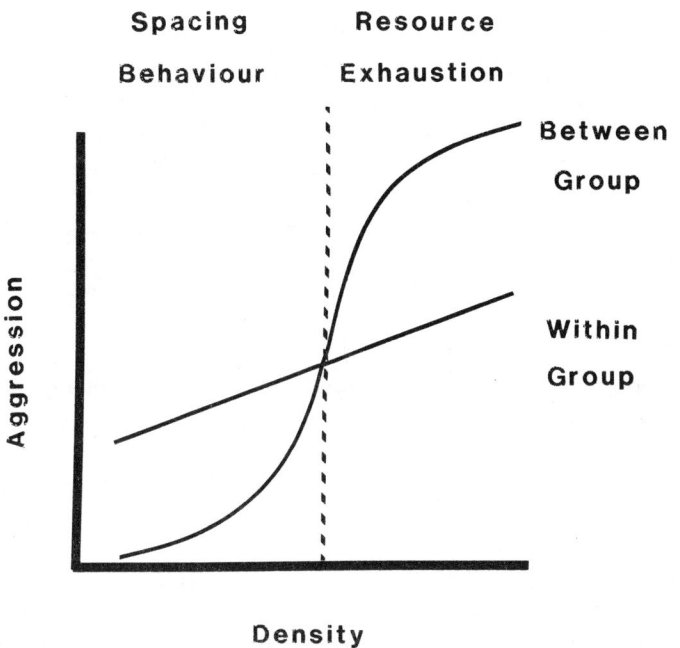

tance at high density in microtine populations (Stenseth 1983).

Although these data are consistent with the social fence hypothesis, the argument is weakened by the inability of food addition to delay declines indefinitely (Taitt and Krebs 1985). In fact, experimental food supplementation appears to be better at accelerating the increase phase of a population cycle, than extending the peak or deferring the decline in voles (Cole and Batzli 1978; Taitt and Krebs 1981; Desy and Thompson 1983; Batzli 1985), snowshoe hare (Boutin 1984a) and grouse (Miller, Watson and Jenkins 1970; Watson, Moss, Phillips and Parr 1977; Watson, Moss and Parr 1984), in direct contrast to what is probably the only unique explicit prediction of the Social Fence hypothesis.

Summary and Prospectus of Argument

In summary, I believe that current hypotheses which seek to integrate population regulation and social behaviour in fluctuating populations of vertebrates are at best weakly supported by available evidence. It is scarcely surprising that the confidence evident in early synthetic reviews of this literature (e.g. Krebs and

Myers 1974), has been replaced by the more quizzical tone evident in the quote from Taitt and Krebs (1985) provided above. Despite this pessimism, I hope my discussion has illustrated that research into these hypotheses has generated a mass of information on social behaviour which does warrant explanation and further investigation. As in other scientific endeavour, peripheral information gathered in attempts to support or reject a hypothesis can often open new vistas of fascinating complexity. From this preliminary review, several such glimpses should be apparent, and these form the basis for the remainder of this text. In particular, I try to understand variation in social behaviour and life history tactics of fluctuating populations in light of new developments in behavioural ecology and see how this interpretation influences an understanding of population regulation.

Many of the hypotheses of population regulation attribute importance to the observation that at low densities microtines are often confined to small isolated patches of habitat. This imposes a population structure which influences both interpretation of evolutionary processes and predetermines the individuals with which animals may interact. In Chapter 2, I address the causes of the patchiness of the habitats in which microtines and other cyclic vertebrates dwell, as a prelude for subsequent discussion of the implications of habitat heterogeneity for the interpretation of social behaviour and life history evolution.

While it was believed that intrapopulation variation in spacing behaviours directly influence population fluctuations, it was anticipated that the phase of a population cycle would influence social behaviour more strongly than would the separate evolutionary histories of species which exhibit similar demography. In Chapter 3, I introduce the complexity of the mating tactics within the microtines. This precedes an analysis of the puzzling diversity of spacing behaviour and an examination of the relations between territoriality, dominance and mating system (Chapter 4).

Although it now seems unlikely that social stress is a major proximate cause of mortality in most populations, observations on the effect of stress indicate that endocrine modulation of reproduction in response to the social environment may influence both social behaviour and life history tactics. In Chapter 5, I analyse the relevance to natural populations of laboratory studies which suggest that social suppression and modulation of reproduction is common in rodents and contrast the evidence supporting this view with adaptive interpretations of variation in time of maturity.

In Chapter 6, the implications of habitat heterogeneity and population fluctuations for the evolution of life history tactics are considered. There is widespread consensus that patterns of emigration and immigration are of fundamental importance in population regulation and social organisation, either as the principal force restricting the growth of populations, or as a fundamental cause of genetic changes within populations. In Chapter 7, I examine the special topic of dispersal and consider hypotheses which suggest that dispersal represents an adaptive response by the disperser, and not emigration enforced by socially dominant residents.

INTRODUCTION 27

Finally, in Chapter 8, the implications of these analyses for the study of population regulation are analysed. Because the data relevant to these themes have often been gathered using methods which are unsuited to distinguishing between the hypotheses I believe are most critical, much of this exercise is intended to be both speculative and stimulating. Wherever possible, I propose tests or field procedures which should provide data to answer the central question - to what extent does demography influence the evolution of social behaviour and life history patterns?

Chapter 2
Habitat Heterogeneity

All animals live in a more or less restricted range of environments. While both chance and inherent aggregation behaviour may lead to restricted dispersion (Roughgarden 1977; Taylor and Taylor 1979), special requirements for habitat characteristics are likely to be the major cause of the patchy distributions of higher vertebrates. An understanding of these habitat requirements is crucial for the analysis of social behaviour and life history attributes, as there is no aspect of the theory of genetics, behaviour, life history evolution or population dynamics in which spatial dispersion is not implicated. In this chapter, I examine the causes of restricted dispersion of the three groups of herbivorous mammals and birds which fluctuate in abundance in a predictable and dramatic manner. The emphasis on herbivory reflects the observation that no carnivorous or insectivorous species has been shown to cycle independently of its herbivore prey, except for species of shrew which are preyed upon by animals which are feeding on much more abundant herbivorous rodents at the same time (Hentonnen 1985; Korpimäki 1986; but see Lomolino 1984). I emphasise the dispersion of food resources, as a prelude to discussion of the implications of habitat heterogeneity for the evolution of social behaviour and life histories. In order to do this it is necessary to introduce systems which attempt to classify habitats according to the selective milieu they generate and briefly discuss the population consequences of **habitat heterogeneity**.

Food

Diet of Voles and Lemmings.

Voles and lemmings of the murid subfamily Microtinae are the chief actors in this account. They occur throughout North America and through much of Eurasia. All of the 110 species of microtine rodents are herbivores, but there is substantial intraspecific and interspecific variation in the the types of plants they consume. While the best known species have generalist diets, others specialise on particular plant groups, or even species. For example, *Myopus schisticolor* appears to eat mosses to the exclusion of other diet items (Bondrup-Nielsen, pers. comm.), and *Phenacomys longicaudus* specialises on the needle leaves of conifers (Hamilton 1962). Analysis of food preferences in other species with more generalised diets is confounded by many methodological difficulties, and the interpretation of adaptation to diet is hindered by the frequency with which studies have been conducted in habitats which are grossly perturbed by human activity. For example, in my own study of habitat use by *Microtus californicus*, 72 per cent of the plant species available to the animal had been introduced to North America to improve pasture quality, or had spread as a result of accidental introduction (Cockburn and Lidicker 1983). Despite these difficulties, a number of impressive attempts have been made to quantify diet and food selectivity by microtines. Although nutrition and the consequences of diversity remain poorly known, recent excellent reviews have identified a number of patterns and unanswered questions which require attention here.

In an analysis of space use and territoriality, Ostfeld (1985a) suggested that there are three biologically important trophic categories which could be distinguished according to seasonality and patterns of food renewal:
(1) Omnivores which feed on fruits, seed, foliage and fungus. Omnivory should reflect seasonal changes in the availability of food items, with specialisation on high quality items during the period they are available;
(2) Forb-eaters which rely predominantly on soft-stemmed dicotyledons or require these plants for health. Forbs are often killed by herbivores and their renewal is likely to be dependent on germination of seeds, which is also likely to be dependent on seasonality. Renewability of the resource following grazing should thus be low; and
(3) Grass-eaters, which have diets comprising at least 60 per cent monocotyledons or other rapidly renewable plant tissue, such as *Equisetum*. Grasses and sedges possess a number of characteristics which ensure high renewability. They can be eaten without being killed because they possess basal intercalary meristems which are inaccessible to damage and facilitate rapid compensatory growth. There has even been speculation that grasses benefit from grazing, as some species disappear or are reduced in abundance in the absence of grazing (e.g. McNaughton 1979; Owen 1980), and the capacity of both sedges and grasses to withstand defoliation is extraordinary (e.g. Archer and Tieszen 1983).

Table 2.1: Dietary characteristics of microtine species

Species	Breeding	Non-breeding	Source
Arvicola			
terrestris	Grass	Grass	Holisová (1965, 1970, 1975, 1976)
Clethrionomys			
gapperi	Seed, fruit		Vickery (1979)
glareolus	Omnivore	Omnivore	Hansson (1971)
rufocanus	Forbs	Forbs	Hansson (1985a)
rutilus	Seeds, Fruit	Seeds, Fruit	West (1982)
Dicrostonyx			
torquatus	Forbs	Forbs	Batzli and Jung (1980)
			Batzli and Pitelka (1983)
Lemmus			
lemmus	Moss, grass	Moss, grass	Kalela *et al.* (1961)
sibiricus	Grass	Grass, moss	Batzli (1975)
			Batzli and Jung (1980)
			Batzli and Pitelka (1983)
Microtus			
agrestis	Grass, forbs	Grass	Hansson (1971)
arvalis	Forbs	Grass	Yu *et al.* (1980)
	Grass	Grass	Holisova (1975)
breweri	Grass	Grass	Rothstein and Tamarin (1977)
			Goldberg *et al.* (1980)
californicus	Grass	Seeds	Batzli and Pitelka (1971)
		Lidicker (1980)	
ochrogaster	Omnivore	Omnivore	Cole and Batzli (1979)
oeconomus	Grass	Grass	Tast (1974)
			Batzli and Jung (1980)
pennsylvanicus	Forbs	Omnivore	Lindroth and Batzli (1984)
richardsoni	Forbs	Forbs	Anderson *et al.* (1976)
xanthognathus	Horsetails	Horsetails	Wolff and Lidicker (1980)
Myopus			
schisticolor	Mosses	Mosses	Tishkov *et al.* (1978)
			Bondrup-Nielsen (pers. comm.)

Arbitrary classifications tend to underestimate intraspecific trophic diversity, particularly that caused by seasonal changes. For example, Ostfeld (1985a) called *M. californicus* a grass-eater, though during the summer drought faced by this species it is almost entirely granivorous in some areas (Batzli and Pitelka 1971). Similarly, *M. arvalis*, which was classified by Ostfeld as a forb-eater, eats little else but grasses during the winter (Yu, Vergne and Gounot 1980) and a variety of foods at other times (Holisova 1959). In Table 2.1, I have attempted to characterise the seasonal diet of microtine species whose population biology is well known, placing emphasis on diet during the periods of reproduction and reproductive quiescence. In formulating this Table, I have relied heavily on the synthesis of Batzli (1985) for American species, and on the best published information for other species. In general, I have preferred information based on stomach contents to that based on faecal analysis. It should be clear that abrupt distinctions between the categories proposed by Ostfeld (1985a) are likely to be more useful in categorising the breeding diets of microtines than in arbitrary assignment of individual species.

Evidence that the diet of many species changes seasonally draws attention to the overwhelming importance of seasonality as a source of temporal heterogeneity to microtines, which are widely distributed at both extreme latitudes and altitudes. In addition to the gross distinctions between winter and summer, a number of authors have reported midsummer nutritional crises in the grass eating *Microtus agrestis* (Myllymäki 1977a), the omnivorous *Microtus ochrogaster* (Rose and Gaines 1978) and fruit eating *Clethrionomys rutilus* (West 1982). While change in diet represents one solution to the general problems of seasonality, some species of both lemmings and voles also exhibit marked seasonal shifts in habitat (Tast 1966; Kalela *et al.* 1971; Batzli, Pitelka and Cameron 1983).

Any change in diet is likely to alter the energetic demands on an organism. Although highly adapted for herbivory, microtines and other small grazing rodents have proved puzzling to physiologists who have studied digestion and nutrition in other mammalian herbivores, particularly domesticated ruminants. In an exhaustive analysis, Demment and Van Soest (1985) consider the capacity of mammals which differ in body size to process plant material. Following Demment (1983), they show that simple kinetic models of the interaction of metabolism and gut capacity suggest that retention time of a food particle (T_r in hours) for mammals exhibiting both fermentation systems scales allometrically according to the relation

$$T_r = 0.589 \, D \, M^{0.28} \qquad \ldots 2.1$$

where D is digestibility of the diet and M is the body mass in kilograms. Thus for a given digestibility, foods will be retained only very briefly by small mammals, creating problems in retaining digesta for sufficiently long to permit extraction of nutrients and energy. This problem is exaggerated in small herbivores as a consequence of their fibrous diet and the high proportion of the diet which is composed of indigestible structural carbohydrates. While some small folivorous mammals exhibit extreme adaptation for energy conservation (e.g. *Neotoma stephensi*, Vaughan and Czaplewski 1985), microtines have both a high rate of metabolism and a high rate of reproduction (e.g. McNab 1980), though not to the extent that was originally believed (Wunder 1985). The difficulties posed by this diet have been circumvented in several ways:

(1) through development of strong food preferences, presumably to increase digestibility (Batzli 1985);

(2) through evolution of extremely efficient dentition, which facilitates comminution of food into very small particles (e.g. Vorontsov 1960; Guthrie 1971);

(3) modification of the proximal colon and caecum (e.g. Snipes 1979). In

particular, the proximal colon of microtines contains an apparently unique mechanism which separates most of the bacteria in the colonic contents from large food particles and returns them to the caecum (Sperber, Björnhag and Ridderstråle 1983). This has the effect of establishing sharp nitrogen gradients from the caecum to the colon and permits maintenance of the colonies of bacteria essential for rapid digestion of small food particles, while allowing the unimpeded passage of large indigestible fragments;

(4) by continuous activity. While omnivorous rodents are often nocturnal, small herbivores commonly feed throughout the day, keeping their guts full at all times (Hansson 1971; Madison 1985);

(5) through coprophagy. Kenagy and Hoyt (1980) showed that *Microtus californicus* reingest about 25 per cent of its faeces and, in association with continuous activity, exhibited a series of rhythmic, short-term alterations between reingestion and non-ingestion during the course of the day and night. Coprophagy is probably common in microtines (e.g. Ouellette and Heisinger 1980);

(6) through adjustment of somatic investment by increasing the length of the gut relative to other tissues. Increase in the length of the gut during periods when the digestibility of the diet declines is known for microtines, hares and grouse (Pendergast and Boag 1973; Moss 1983; Pulliainen and Tunkkari 1983; Gross, Wang and Wunder 1985; Hansson 1985b). Reduction of the gut in conditions of comparative plenty suggests that its maintenance may be costly.

Effects of food on population size.
While these traits afford excellent evidence that food characters influenced the evolution of microtines, demonstration of a proximate effect of food has been more problematic. If these animals are indeed subject to severe food limitations, it seems reasonable to expect that supplementation of food, or increase in its quality, might lead to increased population densities over those ordinarily observed. Although early results were not encouraging (Krebs and DeLong 1966), nearly all experimental studies which have attempted food supplementation have produced substantial demographic effects (Flowerdew 1972; Andrzejewski 1975; Cole and Batzli 1978; Gilbert and Krebs 1981; Taitt 1981; Taitt and Krebs 1981, 1983; Desy and Thompson 1983; Ford and Pitelka 1984). Indeed, Gilbert and Krebs (1981) have proposed that this result is sufficiently general to enable the use of doubling of population density as an appropriate model against which aberrant responses to food (no increase, a dramatic increase) can be compared. They correctly point out that limits on the extent of food-induced increase suggest that food is rarely the only factor limiting increase in population density.

A major problem affecting attempts to relate research on nutrition to research on population regulation and habitat heterogeneity has been underestimation of the complex nature of herbivore diets. First, there is no reason to assume that energy is an appropriate currency for assessing diet selection by herbivores (Pulliam 1975; Westoby 1978). The effects of mineral nutrients and chemicals used by

Figure 2.1: Mean densities of each sex presented for three different habitat types used by Microtus californicus. Agrostis *habitat is dominated by* A. exarata *and several herbaceous dicotyledons.* Holcus *habitat is dominated by* H. lanatus. *Mixed annual habitat is dominated by* Bromus mollis, B. carinatus, B. diandrus *and* Lolium multiflorum. *Density of female voles (stippled bars) is affected by habitat, but density of male voles is not. After Ostfeld and Klosterman (1986)*

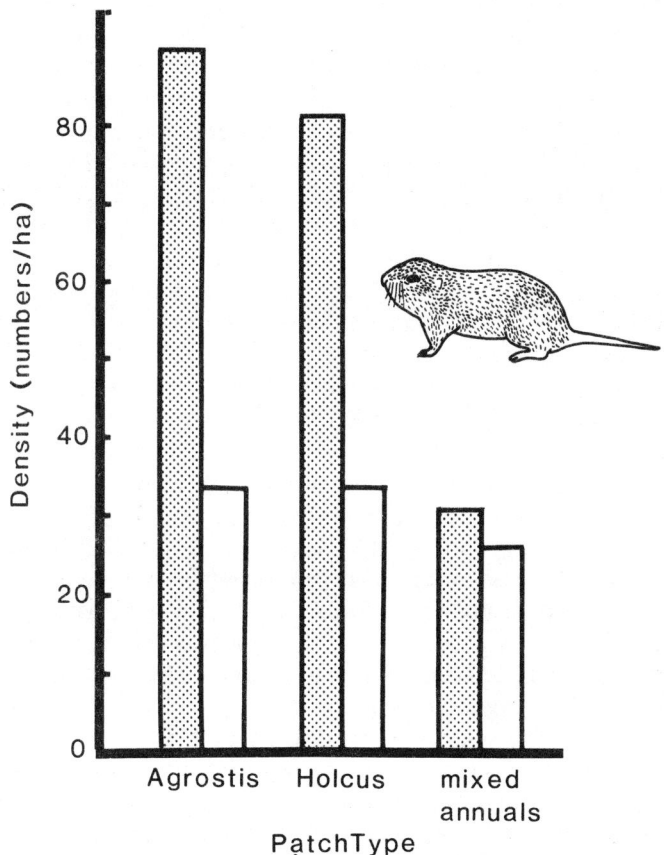

plants as deterrents to herbivory need to be included in an assessment of diet selectivity.

Several studies of diet of *Microtus californicus* can be used to illustrate the complexity of these relations. In an intriguing and poorly understood result, Ostfeld and Klosterman (1986) fed captive voles diets which attempted to simulate the plant species to which the animals were exposed in different habitat patches in the field. All plants in 1 m^2 areas were clipped and frozen, and then supplied *ad libitum* to the captive animals. Although voles grew well on diets dominated by the native perennial grass *Agrostis exarata*, or the introduced perennial *Holcus lanatus*, voles were surprisingly unable to maintain weight on a diet consisting of several annual grasses (*Bromus* spp. and *Lolium multiflorum*) and a variety of small herbaceous dicotyledons. Variation in the ability to maintain weight on different diets was reflected in the abundance and reproductive performance of the voles in different microhabitats in the field (Figure 2.1). These results are in stark contrast to the analysis by Batzli and Cole (1979), who found that *M. californicus* could maintain weight and extract 50 per cent of the energy from a diet of *Bromus rigidus*. Cockburn and Lidicker (1983) showed that exclusion of *M. californicus* from small plots prevented the elimination of many small dicotyledons and enhanced survival and growth of several grasses, including *Bromus mollis* and *Lolium multiflorum*. None of the field derived diets in Ostfeld and Klosterman's (1986) study enabled voles to grow as rapidly as did a diet of fresh carrots and rat and guinea pig chow. However, Krohne (1980, 1981) observed that *M. californicus* in a perennial grassland had smaller litters than laboratory colonies, while voles from annual grassland had larger litters than animals maintained on lab chow. These confusing results raise the possibility that animals exhibit local adaptation to the food plants available to them, confounding the use of comparative tests. Such analyses also provide compelling evidence that subtle floristic variation at a microspatial scale can profoundly influence the dispersion of microtines (Cockburn and Lidicker 1983).

Such spatial differences may confound interpretation in subtle ways. For example, there has been theoretical and empirical speculation concerning the influence of plant deterrents to herbivory on population dynamics of voles (Freeland 1974; Batzli and Pitelka 1975; Haukioja and Hakala 1975; Haukioja 1980; Batzli 1985). Because voles at high densities should be confronted with a reduced opportunity to select non-toxic foods, and should impose natural selection favouring the expression or manufacture of deterrents to herbivory, more toxic foods should be consumed at this time, inhibiting reproduction. Although the connection between toxicity of plants and population dynamics is unconvincing (e.g. Jonasson *et al.* 1986; Lindroth *et al.* 1986), deterrents to herbivory may be relevant in a discussion of spatial heterogeneity. It has often been predicted that there should be variation in the production of herbivore deterrents both within the different tissues of plants and according to the age of the plant, and according to the successional stage to which plants are adapted (e.g. Cates and Orians 1975; Feeny 1975, 1976; Rhoades and Cates 1976; Fox 1981). Under these conditions there may be substantial between-patch differences in the levels of plant defences to which herbivorous vertebrates are exposed. Hence some components of a vole population may be more deleteriously affected than those able to occupy habitat in which degeneration of food quality is less pronounced, or may be forced into habitats where toxic plants are prevalent. Bergeron (1980) was able to demonstrate that *Microtus pennsylvanicus* discriminate between plants according to their potential toxicity. Stomach contents of voles collected during the peak and increase phases of a fluctuation failed to show unequivocally that the abundance of toxic plants eaten by the population as a whole increased with density, but there was a strong effect among subadults (see also Bergeron 1984). Unfortunately no data were presented which allowed the spatial basis of these differences to be investigated.

Despite this complexity, the evidence that food dispersion influences the spatial dispersion of voles is convincing. When more attention is paid to the dispersion of different sex and age classes within a population, these correlations should be strengthened.

Diet of Snowshoe Hare

Unlike many other hares, snowshoe hares (*Lepus americanus*) and arctic hares (*L. timidus*) prefer forested country in northern Eurasia and North America. Both species exhibit cycles of about ten years duration in at least part of their range (e.g. Keith 1963; Keith and Windberg 1978; Finerty 1980). They are browsers, feeding on twigs and leaves from a variety of shrubs and trees. Preferences for particular species, and particular plant parts are well known (see review by Bryant and Kuropat 1980). Whether these reflect avoidance of plant parts containing high concentrations of chemical deterrents, availability of energy or availability of specific limiting nutrients like sodium is not completely resolved, though preliminary data support a role for all three factors (Pease, Vowles and Keith 1979; Moss and Miller 1976; Bryant 1981; Bryant *et al.* 1983, 1985, Belovsky 1984; Fox and Bryant 1984; Pehrson 1981; Sinclair *et al.* 1982). Of particular interest is the inability of hares to survive on a diet consisting of a single species of plant (Keith, Cary, Rongstad and Brittingham 1985), suggesting diet choice reflects several factors.

Demographic effects.

There is substantial non-experimental evidence that food quality declines as hare density increases (Keith 1974; Moss and Miller 1976; Keith and Windberg 1978; Sinclair *et al.* 1982; Fox and Bryant 1984; Keith *et al.* 1985; but see Moss and Hewson 1985), and theoretical reasons to believe that the demographic impact of these changes could be substantial (Belovsky 1984). The first major attempts to halt a hare decline by food addition produced equivocal results (Boutin 1984a; J.N.M. Smith, A.R.E. Sinclair, C.J. Krebs, pers. comms.), which is scarcely surprising given the extraordinary logistical difficulties in carrying out these experiments, and the complex nature of hare diets.

Spatial heterogeneity.

As for voles, it is easy to accept that there might be substantial temporal heterogeneity in browse availability to hares (Pease *et al.* 1979), but the spatial aspects of hare dispersion have only recently been investigated. Because there appears to be a close and undisputed relation between the abundance of hares and lynxes, it is possible to infer hare density from data on the gathering of lynx pelts by fur trappers over the last two centuries (e.g. Elton and Nicholson 1942; Finerty 1980). These long time series facilitate the use of spectral analysis to identify the phase relations of a cycle, and this technique has been used by Smith and Davis (1981) to examine the temporal relations between population changes of snowshoe hares in different regions in Canada. Their analysis confirms a widespread suspicion that an increase or decline is initiated in a nodal region in the west of Canada and then spreads towards the extremes of the country over a period of two to four years. They also suggest that there is a drift in the position of the node between cycles. These large scale changes are of interest, but obviously are unlikely to be perceived by animals whose vagility restricts them to movements over a much smaller scale.

Wolff (1980) demonstrated that seasonal movements between habitats were an important aspect of diet optimisation in a cyclic population of hares in Alaska (see also Pulliainen 1982, 1983). Where habitat refugia such as dense stands of spruce or thickets of willow and alder are available, animals feed during summer in open areas where they consume forbs and browse from low shrubs. In winter this food is rendered inaccessible by snow and the animals move into thickets where they browse on spruce (*Picea*), willow (*Salix* spp.) and alder (*Alnus* spp.). These refugia not only provide winter browse but can act as refugia from predation, which intensifies during a hare decline and the subsequent low phase. Hares which spread to other habitats during a population peak are exposed to predation, and as a consequence the population ultimately becomes confined to small refugia. Fox (1978) argues that post-fire succession will be important in determining the availability of suitable habitat, and even that fire patterns are capable of driving the ten year cycle.

Diet of Grouse and Ptarmigan

The 16 species of grouse and ptarmigan are gallinaceous birds in the subfamily Tetroninae, and have a circumpolar distribution throughout North America and Eurasia. The extent of cyclicity is variable. Rather than attempt a complete review of tetraonid diets, I shall briefly summarise the exemplary analysis of the importance of food in Scottish populations of the red grouse, *Lagopus lagopus scoticus*. Moss and Watson (1985) provide an excellent review of this work. Although grouse may occasionally feed on invertebrates, their diet is dominated by browse from an evergreen ericoid shrub, *Calluna vulgaris*. Grouse feed selectively both between and within patches of heather, and it has been possible to establish simple correlations between this preference and nitrogen availability (Moss 1972; Lance 1983; Savory 1983), though this is modified to some extent by preference for easy access to food (Moss, Miller and Allen 1972). This preference has long allowed management of moors by burning to stimulate post-fire succession and consequently enhance bags of grouse, and prompted ecological experiments which confirm the validity of both this technique (Miller, Watson and Jenkins 1970), and heather stimulation by the addition of fertiliser (Watson and O'Hare 1979). However, while food supplementation enhances growth of populations and survival and reproduction of hens, it has not yet been possible to defer a decline indefinitely (Watson and Moss 1980; Watson, Moss and Parr 1984; Moss and Watson 1985).

Other Factors

Although this discussion has placed particular emphasis on the role of dietary factors in influencing dispersion of microtines and other vertebrate herbivores which fluctuate in abundance, it is important to recognise that distribution and abundance are influenced by other extrinsic factors, of which the chief are the impacts of predators and disease, the hostility of the climate and competition with other species.

Predation

There is little doubt that many voles end their lives in the stomach of a predator, but the impact of predation on microtine populations remains controversial. Descriptive studies inevitably report high incidence of predation (e.g. Pearson 1971, 1985) but their methodology is subject to many criticisms. For example, Pearson (1985) used the number of vole remains found in carnivore pellets to estimate that predation accounted for 88 per cent of a *Microtus californicus* population. However, he was unable to distinguish between true predation and consumption of carrion (see Mullen and Pitelka 1972), and was forced to rely on runway density and trapping indices to estimate the number of voles available to predators. This index is likely to underestimate both dispersers and juveniles (Boonstra and Krebs 1978; Beacham and Krebs 1980; Boonstra and Rodd 1984), which may be more susceptible to predation (see Lidicker 1975 for a review).

Fennoscandian authors have also proposed that the gradient in cyclicity with latitude is indicative of an influence of predation. In southern Fennoscandia the predators of microtines are diverse, and many have access to a range of alternative prey, and consequently should exert a stabilising influence on microtine populations. By contrast, at northern latitudes, predators are more specialised and consequently more likely to drive microtines to very low densities. In particular, the absence of alternative prey should force the predators to search for microtines at very low densities, prolonging the period before populations of small mammals can recover (Andersson and Erlinge 1977; Fitzgerald 1977; Erlinge *et al.* 1983, 1984). A similar argument has been advanced to account for geographic variation in the extent of cyclicity in populations of snowshoe hare (Wolff 1980). The abundance of other small mammal species may also be driven by these fluctuations (Hörnfeldt 1978; Angelstam *et al.* 1985; Hentonnen 1985; Korpimäki 1986). Korpimäki (1985) has argued that some species of vole specialists may stabilise numbers in the same way as generalists. These species (owls and raptors) are nomadic and exhibit a dramatic response to microtine irruptions. Because the nomadism ensures that there is very little time lag between an irruption of voles and an irruption of owls, the predators have the potential to truncate peaks but not sustain the low phase. Theoretical analyses suggest that the relations between time lag and population stability are rather complex, and warrant further investigation (Hastings 1983a; Nunney 1985b).

The best experimental work on predation on microtines formed part of a multifactorial analysis of the tendency of *Microtus townsendii* to decline during the spring. Taitt and Krebs (1983) showed that predation by great blue herons (*Ardea herodias*) was likely to depress numbers during the early part of spring when cover was poor but would be of minor importance once vegetation regeneration improved cover. Experimental research on the impact of specialist vole predators such as small mustelids has not been attempted, and nor has there been extension of any experimental research into the low phase of a population to which descriptive accounts attribute particular importance. Nonetheless, the results of Taitt and Krebs (1983) are encouraging as they suggest it may be possible to distinguish the effects of food availability, cover and predation, between which some spatial correlation might be anticipated. This may allow experimental dissection of hypotheses concerning the importance of predation and cover (see e.g. Wolff 1980; Birney, Grant and Baird 1976).

It is worth pointing out that an inadvertent set of experiments may have been carried out over many years by applied biologists attempting to control rodent populations. The overwhelming failure of trapping and poisoning to accelerate the decline or to defer indefinitely the resurgence of rodent populations may represent good experimental evidence that predation is not a major factor in population regulation. As Elton's famous quote attests (1942, p. 3):

> The affair runs always along a similar course. Voles multiply. Destruction reigns. There is dismay, followed by outcry, and demands to Authority. Authority remembers its experts or appoints some: they ought to know. The experts advise a Cure. The Cure can be anything : golden mice, holy water from Mecca, a Government Commission, a culture of bacteria, poison, prayers denunciatory or tactful, a new god, a trap, a Pied Piper. The Cures have only one thing in common : with a little patience *they always work*. They have never been known entirely to fail. Likewise they have never been known to prevent the next outbreak.

However, the implication of the assertion that predation is capable of maintaining low densities but not of initiating a decline is clear. Managers should focus their efforts on the periods between peaks, instead of responding after the horse (or vole) has bolted.

Parasites and Disease

Although the impact of parasites and disease on small mammal populations proved attractive to workers concerned with control (see Elton 1942 for a review), the impact of disease remains unresolved. In Chapter 1, I discussed evidence that the proximate cause of mortality in some circumstances may be a consequence of opportunistic invasions or flourishing of parasites following suppression of the immune and inflammatory systems in response to stress. This observation raises the spectre of circularity. Does a negative correlation between survival or reproduction and the parasite burden indicate that parasites are a principal contributor to reduced fitness, or are weaker animals more susceptible to parasitism, or do animals trade increased parasite loads for the advantage of tissue mobilisation through gluconeogenesis? A research programme to distinguish between these three alternatives in the field has not been developed. For example, Hudson (1986) was able to demonstrate a negative correlation between nematode infection and reproductive success in red grouse both between and within habitats. Although experimental treatment with an anthelmintic increased reproductive success, confirming that parasite loads rather than between bird and patch differences are directly responsible for the reduction in fitness, the ultimate causes of susceptibility to parasites remains unexplored.

Possible causes of spatial heterogeneity in susceptibility to disease are easy to understand, but remain poorly investigated. Local infection rates will increase as host density and contagion increase, so any extrinsic or intrinsic factor which causes patchy distribution of animals will in turn influence the distribution of their pathogens. Modelling of spatial distribution of parasitism and disease needs to take into account the dispersion of pathogens both between and within hosts (Anderson and May 1978; May and Anderson 1978), but relevant data are unavailable from the vertebrate taxa of interest.

Climate

The frequency with which periodic irruptions occur among vertebrates living in high arctic latitudes is evidence that climate may exert an influence on cyclicity (Finerty 1980). Kalela (1962) and Tast and Kalela (1971) have hypothesised that the short growing season of plants at northern latitudes produces an endogenous rhythm in flowering phenology which influences fluctuations of small mammals in the tundra. Multiannual synchrony in flowering and seed production between plant species might be initiated by heavy browsing during a microtine high. Variation in growth conditions may influence the number of years required for flowering and consequently the interval between peaks. Although other Finnish researchers have argued in support of this hypothesis (e.g. Laine and Hentonnen 1983; Hentonnen et al. 1984; Tast 1984), there is currently little evidence supporting its importance (Hansson 1979b; 1984a; Hentonnen 1985; Andersson and Jonasson 1986; Hörnfeldt, Löfgren and Carlsson 1986). A similar argument was used by Pitelka (1964) and Schultz (1964) who suggested that recovery of plants from overgrazing would take several years, but even the chief advocates of this hypothesis are now sceptical about its generality (Batzli et al. 1980; Batzli 1985), even though there is good evidence that grazing by rodents influences plant biomass at extreme latitudes (Oksanen 1983).

Competition

Voles are usually the dominant small mammal species in the communities in which they occur. For example, *M. californicus* rapidly displaces other rodent species when it attains high densities in grassland (Blaustein 1980, 1981; Heske, Ostfeld and Lidicker 1984). However, because many species of microtines occur parapatrically or sympatrically, there is also potential for competition between vole species. The occurrence of such competition is supported by both observations of habitat segregation (e.g. Grant 1975, Randall 1978; Hansson 1982 and many others), and removal experiments (e.g. Linzey 1984, but see Gilbert and Krebs 1984). There is no evidence that competition between vole species is a driving force in either social evolution or population dynamics. Indeed, the synchrony of irruptions in some sympatric species suggests that interspecific competition can be ignored in the development of my argument.

Habitat Use

The significance of habitat heterogeneity for the biology of fluctuating populations is widely acknowledged (Hansson 1977; Anderson 1980; Rosenzweig and Abramsky 1980; Stenseth 1980; Wolff 1980; Cockburn and Lidicker 1983; Lidicker 1985a; Taitt and Krebs 1985 and many others). It is less clear what are the best means to identify habitat preference, or what are the best criteria with which to assess habitat suitability for small mammals (Cockburn 1984). Taitt and Krebs (1985, p. 593) comment: 'There is no agreement about the role of chance in spatial heterogeneity, and this has led to circularity. Do we distinguish optimal habitats by their vegetation characteristics or by the fact that they always contain voles?' In addition, Van Horne (1983) and Cockburn and Lidicker (1983) have pointed out that there need not be a simple positive correlation between density of animals and aspects of habitat which facilitate survival and reproduction. Van Horne (1983) identified six circumstances which might cause this decoupling:

(1) where habitat availability varies seasonally, density is unlikely to be a reliable indicator of quality when only a single season is sampled;

(2) unpredictability through time means that some habitats may be exploited for short periods but be unable to sustain reproduction and survival for ecologically and evolutionarily important periods;

(3) patchiness will restrict colonisation and survival of populations in some habitats, particularly where a source population is absent;

(4) social interactions are particularly likely to restrict access to high quality habitat, lessening the prospect of a temporary overshoot in such habitat. The density of animals in marginal habitat may therefore temporarily exceed that in high quality habitat;

(5) high reproductive capacity will promote temporary high densities in some habitats; and

(6) these effects will be most pronounced in species capable of exploiting a broad spectrum of habitats (generalists).

In order to illustrate these difficulties, consider the population of *Microtus californicus* studied by Cockburn and Lidicker (1983) and Ostfeld, Lidicker and Heske (1985). In grassland dominated by annual species which wither during the harsh Mediterranean summer in central California, the voles decline to extinction during the summer. By contrast, a few voles survive the summer in perennial grassland, particularly in the summer preceding a peak in abundance, and this perennial grassland is favoured by breeding female voles at all times. The high survival can be related directly to the water requirements of the voles (Lidicker 1976; Nelson, Dark and Zucker 1983). Reproduction commences following rains in the winter, and unoccupied habitat is colonised as voles disperse into the annual grassland. Following an additional spring flush in growth of grasses, vole density in the annual grassland may temporarily exceed that in the perennial grassland, contrary to the expectation that the preferred vegetation should support most voles (Figure 2.2). The discrepancy arises because the major contributors to

Figure 2.2: Number of potentially reproductive female Microtus californicus *resident on two trapping grids for sufficiently long to complete reproduction. One grid (circles) contains a mixture of perennials and annual grasses, while the other (triangles) is dominated by the annual grass* Elymus triticoides. *After Cockburn and Lidicker (1983)*

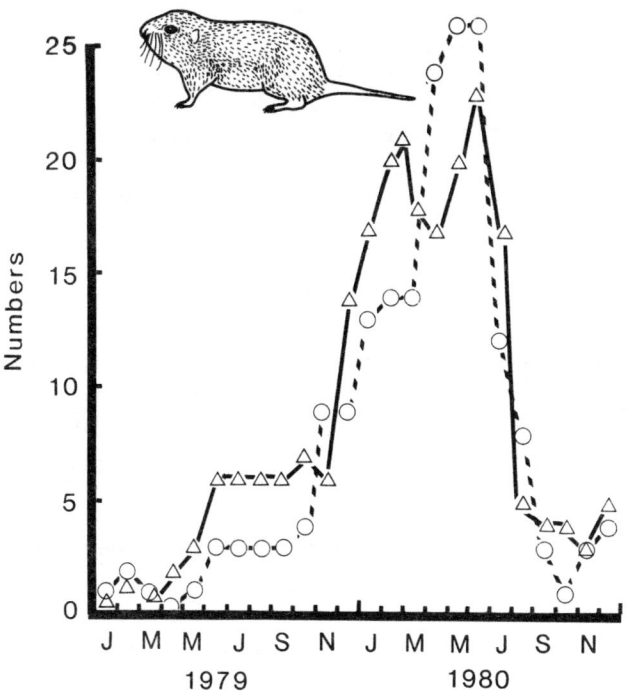

this increase are young males with restricted opportunities for reproduction, as they are excluded by territorial males from high concentrations of reproductive females in the perennial grassland (Ostfeld 1986). Therefore it is possible to recognise the actual or potential contribution of all the problems identified by Van Horne (1983).

The Classification of Habitat

Classification of habitats has also created problems with circularity. It is widely believed that the temporal and spatial characters of habitat profoundly influence the evolution of the life histories of the animals which live within them, prompting several attempts to classify habitats acording to the selective milieu they generate (e.g. Pianka 1970; Wiens 1976). However, the same spatial and temporal properties only have biological relevance when considered with reference to the organisms which inhabit them (Stearns 1976, 1977; Southwood 1977). A temporary pond formed after heavy rain may allow some small invertebrates to complete several generations, but the same pond may dry up before the eggs of a salamander can complete development (Gill 1978b). Are the small invertebrates adapted to life in ephemeral ponds, or does the brevity of their reproductive cycle make the life of the pond irrelevant? The most forceful example of the confusion caused by this circularity has been described by Parry (1981) in his review of the various interpretations of the influential but notorious theory of r- and K- selection, to which I have alluded in Chapter 1. This theory is most usefully applied to consideration of whether populations which live in fluctuating environments which are rarely saturated with conspecifics differ from populations which live in more stable environments where the carrying capacity is usually attained. In other terms, is selection on life history evolution density-dependent? (Boyce 1984). However, Parry points out that at least four separate meanings of r- and K-selection are prevalent in the literature (Table 2.2). Meaning 1 refers to an expected outcome of selection in environments according to the level of crowding. Meaning 2 refers to the density-dependent aspects of natural selection. Meaning 3 refers to an expectation about the species which can be anticipated in different habitats, and Meaning 4 defines the result of selection in terms of life history traits. The empirical and theoretical problems with this model have been well discussed elsewhere (Stearns 1976, 1977; Hart and Begon 1982; Boyce 1984). In particular, it is well known that the appropriate result is sensitive to whether the effects of density on adult or juvenile mortality are stochastic or deterministic, suggesting that the life history traits of the animals cannot be ignored in generating predictions.

In order to overcome the simplicity associated with classification of habitats along a single axis (stable versus fluctuating, or unsaturated versus saturated), several other classifications of habitat characteristics have also appeared in the literature on life history evolution. The first was developed by Grime (1977, 1979) and Greenslade (1983) and attempts to classify habitats according to two

Table 2.2: The meanings of r- and K- selection (modified from Parry 1981)

Meaning	r-selection	K-selection
1.	selection to maximise population growth in uncrowded populations	selection for competitive ability in crowded populations
2.	density-independent component of natural selection	density-dependent component of natural selection
3.	r-species occur in habitats which are ephemeral	K-species occur in habitat with a long durational stability
4.	allocation of a large proportion of resources to reproduction	allocation of a small proportion of resources to reproduction

axes, the intensity of disturbance (Grime) or predictability of the environment (Greenslade) and the intensity of 'stress' (Grime), or favourableness (Greenslade) of the environment (Table 2.3). Three possible outcomes were predicted, with particular emphasis being placed on the outcome of selection in predictable but poor environments, conditions which are not examined by the r-K dichotomy (Table 2.3; Figure 2.3).

A two factor classification was also suggested by Sibly and Calow (1985), who classified habitats according to life history attributes themselves. These are G, an index of the growth rate of offspring, and an index of survivorship, S, the definition of which depends on the nature of the costs incurred in reproduction. Costs could be incurred before eggs are laid (*direct costing*), or after the eggs are laid (*absorption costing*) (Sibly and Calow 1984). Separate formulations for each case were used to derive predictions about three aspects of reproductive investment (the total investment, the investment per egg, and the number of eggs) (Table 2.4). According to this view, some life history characters (growth and survivorship) are a proximate response to habitat and presumably should always be maximised, while others (investment at a given time, and the division of investment among embryos) are an adaptive ultimate response to habitat characters. While this is a commendable attempt to escape circularity, the notion that growth should always (or can be) maximised is controversial (Calder 1984). Nor is it clear whether higher vertebrates sustain costs of reproduction before or after the production of eggs, or that the energetic costs implicated by Sibly and Calow (1984, 1985) are the appropriate currency with which to interpret life history tradeoffs (Chapters 5, 6). Further, the theory currently fails to deal with the effects of predictability, and also with the effects of resource availability on costs of reproduction (Sibly and Calow 1985).

By contrast, Begon (1985) classifies habitats according to the relation between the accumulated somatic investment (ASI) of an organism and its residual reproductive value (RRV), or the sum of future reproductive opportunity devalued by the costs of deferring reproduction (Williams 1966b; Schaffer 1981). Four habitats were recognised according to the effect of accumulated somatic invest-

Table 2.3: The two axis habitat classification systems of Grime (1977, 1979) and Greenslade (1983), and the synonymies they prefer with other systems of life history prediction according to habitat characteristics

(a) Grime Disturbance	'Stress' Low	High
Low	Competitive strategy (= α-strategy)	Tolerant strategy (= K-strategy)
High	Ruderal strategy (= r-strategy)	No viable strategy

(b) Greenslade Predictability	Favourableness High	Low
High	K-selection	Adversity selection
Low	r-selection	r-selection

ment on either the residual reproductive value of established individuals or that of their offspring (Table 2.5). A strong correlation between ASI and RRV could arise in two quite distinct ways. First, large ASI (good growth or condition) may be the best source of protection against hostile extrinsic factors, such as predation or climate. Second, large or old animals may often perform better in competition to survive and reproduce. Lack of correlation or a negative correlation could be generated in three ways. There will be no correlation if mortality is catastrophic and unavoidable, or if conditions are sufficiently benign to render intraspecific competition irrelevant. There will be a negative correlation if mortality affects the largest animals more severely (e.g. when a predator selects large prey). This analysis was used to predict the occurrence of semelparity and allocation of resources among broods in each of the habitats. This is also an attempt to separate proximate and ultimate aspects of the effects of habitat, but Begon (1985) concedes that this is a descriptive demographic model rather than a predictive evolutionary analysis.

The best model which specifically attempts to describe habitat use by small mammals in temporally fluctuating environments had its genesis in Russian research in the middle of the century (see e.g. Elton 1942), and is normally attributed to Naumov (1955, 1964) (but see Fenyuk 1937; Lidicker 1985a). The model has been reviewed on a number of occasions (e.g. Anderson 1970, 1980; Hansson 1977; Myllymaki 1977a; Smith, Manlove and Joule 1978), and distinguishes habitats according to the period they remain suitable for both occupancy and reproduction. Because there is little consistency in the use of terms to describe the different habitats recognised in the Naumov model, I have tried to draw together the varying terms under the general habitat templet proposed by Southwood (1977), as modified by Begon and Mortimer (1981) (see Table 2.6).

Figure 2.3: The habitat templet proposed by Grime (1977, 1979) and Greenslade (1983). Variation in the intensity of the hostility of the environment and the extent of predictability are hypothesised to generate three selective conditions: r-selection, K-selection and adversity (A) selection. The line of biotic unpredictability refers to the increased extent to which distributions are affected by competition with other animals or plants when conditions are favourable and predictable. Modified from Greenslade (1983)

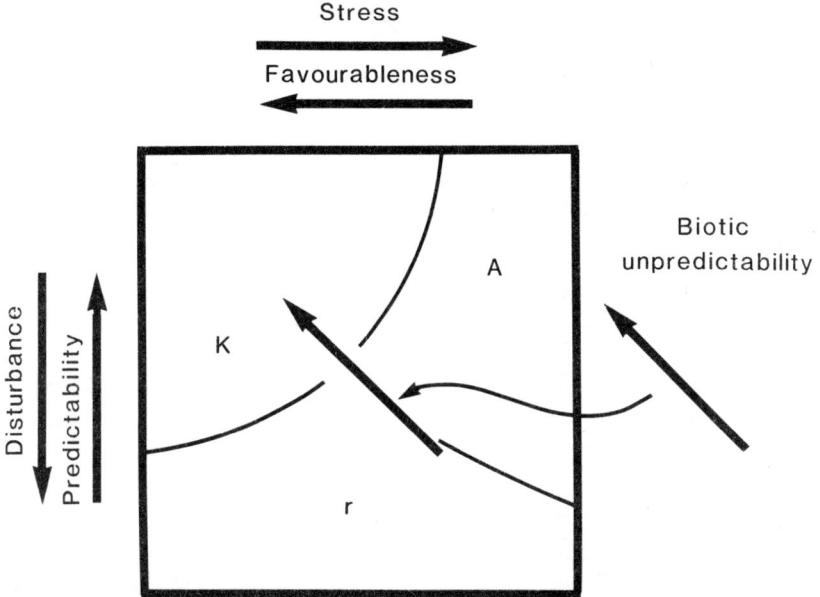

There seems little doubt that populations which fluctuate in abundance are confined for much of the time to small, and often isolated, patches of habitat (Newsome 1969a, 1969b; Wolff 1980; Mazurkiewicz 1981; Cockburn and Lidicker 1983), which provide conditions suitable for year round occupancy (constancy), and occasional (normally seasonal) reproduction. At least some of the population growth during fluctuations appears to be associated with an increase in the range of habitats occupied, particularly during periods adverse for reproduction (Hansson 1979a; Cockburn and Lidicker 1983; see however Batzli, Pitelka and Cameron 1983). A major factor contributing to the extent of the spring irruption into new habitats will be any increase in the period over which reproduction occurs (winter breeding e.g. Krebs 1964a; Myllymäki 1977a; Hansson 1984b; Kaikusalo and Tast 1984), or in the range of habitats which permit survival during the most harsh season.

A pattern of habitat occupancy of this sort has some interesting consequences for interpretation of evolutionary dynamics in fluctuating populations. First, Anderson (1970) has argued (I believe correctly), that the repeated invasion of habitats which are only temporarily available leads to a group of animals which make no long-term genetic contribution, and can therefore be largely ignored in

Table 2.4: Sibly and Calow's (1985) prediction of life history or life cycle traits where habitat characteristics influence two variables, G, an index of the growth rate of offspring, and S, an index of survivorship influenced by the time at which the costs of reproduction are incurred

Z = total investment in reproduction in a given attempt
w = investment per egg or offspring
n = number of eggs or offspring

		Survivorship (S)	
		Low	High
Growth (G)	High	Z low w low n intermediate	Z high w low n very high
	Low	Z low w high n low	Z high w high n intermediate

interpretation of life history evolution (Figure 2.4). As an overwhelming proportion of microtine research focuses on peaks of abundance (when new habitats are being invaded), rather than lows, the data base may be biased in a manner which precludes useful evolutionary analyses. For example, Cockburn and Lidicker (1983) demonstrated a correlation between several floristic attributes and the dispersion and reproductive success of female *Microtus californicus*. The correlation persisted throughout a cycle. They were unable to explain the intense preference in terms of improved habitat during the breeding season, but were able to demonstrate that habitat attributes should help voles survive the summer, or period of greatest stringency in central California. Non-reproductive habitat requirements appear to dictate habitat selection in this species, yet it is unlikely that this would be recognised in a short-term study at high density.

Second, while it is clear that all the classifications of habitat described above deal with the effects of temporal heterogeneity, none deal adequately with the evolutionary consequences of seasonality.

Large-scale temporal heterogeneity.
Superimposed on this seasonal scale of change will be the dependence of many species on environments in early stages of succession (Southwood, Brown and Reader 1983). The requirement of many *Microtus* species for grasslands which will eventually be supplanted by forest or woodland is well known (e.g. Getz 1985), and it is almost certain that many species have increased in abundance and

Table 2.5: Begon's (1985) classification of habitats in terms of their effect on the relation between accumulated somatic investment (ASI) and residual reproductive value (RRV). Although habitats are classified by the influence of this relation on either established individuals or their offspring, the causes of the relation can be expected to be common to both population subclasses

	Established individuals	Offspring of established individuals
RRV increases with ASI (e.g. age, condition)	Type 1	Type 3
RRV is little affected by or decreases with ASI	Type 2	Type 4

distribution as a consequence of the conversion of woodlands to grasslands for agriculture. However, woodland and forest species also exhibit successional dependence. For example, *Microtus xanthognathus* occupies post-fire successional vegetation and persists there for 10 to 20 years (Wolff and Lidicker 1980), and fire may stimulate growth and promote availability of the food species required by *Lepus americanus* (Fox 1978). *Microtus oregoni* utilises habitat generated in post-logging succession (Sullivan 1980; Sullivan and Krebs 1981). Tundra succession may also be profoundly influenced by fire (e.g. Fetcher, Beatty, Mulinaux and Winkler 1984).

Large-scale spatial heterogeneity.
In addition to this broader temporal perspective, it is also useful to consider the broader spatial perspective which may have influenced the evolution of these animals. Getz (1985) has attempted a classification of the type of habitats to which North American *Microtus* species might have been exposed (Table 2.7). It is important to note that there is no relation between patch configuration and distribution. For example, *M. pennsylvanicus* is supposed to have been confined to small isolated patches despite its extremely broad distribution. By contrast, *M. townsendii* was expected to be exposed to stable contiguous habitats within its narrow range. In Eurasia, where alienation of land for agriculture preceded an academic interest in microtine biology, the prospect of extending this analysis is restricted. However, it has been suggested that *M. agrestis* was once an inhabitant of patchily distributed mires, and of openings generated by burning. *Clethrionomys* spp. were probably always more prevalent in forests than *Microtus* spp. (Boström and Hansson 1981; Hansson 1984c).

Summary and Future Directions

The dispersion of microtines, hares and grouse is restricted by the patchiness of

Table 2.6: A classification of habitat suitability for small mammals, modified from Southwood (1977) and Begon and Mortimer (1981), with synonyms provided by previous authors. Note (a) the confusing use of the term survival habitat, and the greater precision in the classifications by Hansson (1977) and Myllymäki (1977a), and (b) the use of R_d for migratory range, to avoid the use of R_m by Begon and Mortimer (1981), with its connotations of similarity to the Malthusian parameter r_m

t — generation time
F — period a habitat remains suitable for breeding
F_{ann} — breeding period within one year
L — period a habitat remains suitable for occupancy (non-breeding)
L_{ann} — non-breeding period within one year
H = F + L
R_f — the foraging or 'trivial' range of an organism
R_d — the migratory range, or dispersal capacity, of an organism
U — the distance between habitat patches, where F = L = 0, but movement is not frustrated
$V_{F,L}$ — variance in the period of habitat suitability

Suggested terms	Constant	Colonisation: Production	Colonisation: Reception	Invasion	Traversable
Anderson (1970, 1980)	Survival (Reservation)	Colonisation	Colonisation	Colonisation	Traversable
Smith et al. (1978)	Primary	Secondary	Secondary	Secondary	—
Myllymäki (1977a)	Primary production	Temporary production	Survival	Invasion	Transition
Hansson (1977)	Donor	Induced donor	Reception	Invasion	Transition
Temporal Axis Non-breeding	Constant	Unpredictable or Seasonal	Unpredictable	?	

Table 2.6 (continued)

Breeding	Seasonal	Unpredictable	Ephemeral or Unpredictable	Ephemeral
Spatial Axis	Patchy or Continuous	Patchy	Patchy	Continuous
Algebraic formulation	$F_{ann} + L_{ann} \approx$ one year	$t < F$	$t > F$	$t < F$ $\quad$ $U < R_d$
	V is low	V is high	V is high	V is extreme

50 HABITAT HETEROGENEITY

Figure 2.4: Implications of seasonal dynamics and habitat heterogeneity for population dynamics in small mammals. Although the number of animals in colonisation habitat may exceed the number of animals in constant habitat, very few colonists will survive the season of greatest stringency except in exceptional years. In these conditions an outbreak may occur. Horizontal bars denote the period colonisation habitat is available. Shading denotes different cohorts, and the width of the bar the number of animals in each cohort. After Anderson (1970)

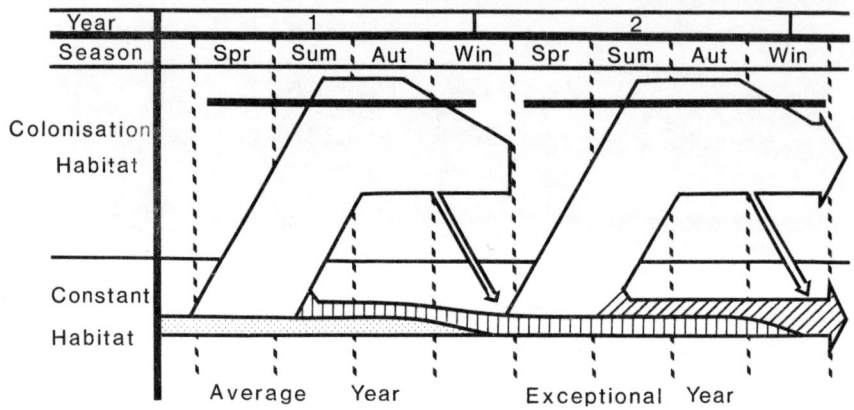

the environments they inhabit, and the causes of this dispersion are crucial to an understanding of the evolution of life histories and social organisation in these taxa. There is currently excellent evidence that the microspatial dispersion of animals is influenced by the availability of food, but experimental evaluation of this evidence is confounded by the subtlety and complexity of these relations. On a temporal level, the hostility of the environment during the non-reproductive period suggests that the impact of seasonality on life history evolution requires careful evaluation. While available systems of habitat classification are appropriate for analysis of the spatial influences on sociality and life histories, seasonal effects are poorly explored in those classifications which attempt to predict the relation between habitat characteristics and life history evolution.

On a broader scale, the temporal pattern of habitat availability is probably dictated by plant succession. In forest and woodland, the frequency and size of fire may exert a similar influence on all the taxa under consideration. In grassland, the selective influences of succession are clouded by the recent alienation of habitat for agriculture, rendering much current research unsuitable for dissecting the selective milieu in which microtines evolved.

It has been suggested that the intensity and characteristics of predation change through the latitudinal range of these taxa, and that because predators do not have alternative prey at northern latitudes, they will continue to harry cyclic populations of vertebrates at low densities, and have a limited capacity to respond quickly to irruptions. This selective predation may influence both population

Table 2.7: Presumed habitat-patch configuration occupied by *Microtus* species prior to human disturbance. Modified from Getz (1985)

Large, contiguous, relatively stable habitats
    M. californicus   *M. oeconomus*
    M. miurus   *M. pinetorum*
    M. ochrogaster   *M. townsendii*

Small, isolated or ephemeral habitat patches
    M. canicaudus   *M. parvulus*
    M. chrotorrhinus   *M. pennsylvanicus*
    M. longicaudus   *M. quasiater*
    M. mexicanus   *M. (Arvicola). richardsoni*
    M. montanus   *M. xanthognathus*
    M. oregoni

Entire species restricted to one or a few very small or localised areas
    M. abbreviatus   *M. coronarius*
    M. breweri (island form)

dynamics, and the nature of the habitat which is the source of irruptions.

Topics for Future Research

(1) Almost nothing is known of the proximate behavioural, or ultimate genetic basis for habitat selection by voles. Wiens (1976) has pointed out that it is important to distinguish between genuine habitat preference and habitat correlations, which might arise in the example given above because animals in some habitat are susceptible to predation rather than because the habitat affords superior food resources. It has been suggested that early experience is critical in determining where *Peromyscus leucopus* settle (Wecker 1963; Barry 1984). The research programme followed in these studies seems amenable to repetition in other species.

(2) Seasonal migration by lemmings and hares is a particular problem. While such migration is easily explained in terms of escape from seasonal adversity, it is not clear whether migrants risk displacement from their breeding by intraspecific competitors or their offspring, or whether short life minimises such competition. Successful seasonal migration would help support the idea that some animals which occupy colonisation and invasion habitat are likely to donate offspring to the constant habitat in non-migratory species. A genealogical perspective on these problems seems most promising.

(3) At a theoretical level the interaction between seasonality and succession as the major determinants of temporal heterogeneity requires further attention. Because the changes in time which accompany succession will not be predictable spatially, but the changes associated with seasonality should be, the availability of patches for colonisation may exert unanticipated selection pressures.

Chapter 3
Mating Options

Although considerable progress has been made in identifying environmental conditions which promote the evolution of different mating strategies, the study of these correlations is currently being reassessed, and involves a more mature and careful assessment of costs and benefits than has previously taken place (Jarman 1982; Vehrencamp and Bradbury 1984). It is clear that mating systems are often evolutionarily labile, but it is still necessary to interpret them within the perspective of phylogenetic constraints. For example, the overwhelming majority of birds (> 90 per cent) mate monogamously, but only about 5 per cent of mammals do. While it has proved illuminating to focus on the evolution of polygynous systems based on resource defence in birds (Verner 1964; Verner and Willson 1966; Orians 1969), there is no particular reason to believe that the same research programme will be useful in the study of mammals. The preponderance of polygyny in mammals may reflect either phylogenetic inertia (Dobson 1985) or the unique way in which female mammals provision their young (Kleiman 1977; Wittenberger and Tilson 1980).

Mating systems among the microtines have proved to be surprisingly diverse, particularly given the emphasis on social behaviour as a unifying thread in interpretation of population regulation in these animals. This diversity and the obvious interest in demography in this taxon suggests that microtines may be useful models for study of the evolution of mating systems, provided the formidable difficulties in studying the behaviour of secretive small mammals in the wild can be overcome. In a thoughtful and provocative review of research on the ecological determinants of mating systems, Vehrencamp and Bradbury (1984) compiled an exhaustive list of the components of fitness which would have to be assessed if the total contribution of behavioural options to the evolution of mating systems were to be understood (Table 3.1). While there is still no comprehensive understanding of the mating system of any microtine species, it will become clear in the next three chapters that investigation of microtine social behaviour has touched on almost all of these components. A brief introductory review of mating options will be helpful in elucidating the direct and indirect approaches currently used in the study of microtine behaviour.

Options Available to Males for Increasing Fitness

Increasing Encounter Rate with the Opposite Sex

The small territories or home ranges of male microtines probably ensure that they are quickly aware of the presence of a nearby female in oestrus. Olfaction is the most likely cue. Where females are territorial, this approach is problematic, but it may be profitable for males to stay near sites where several territorial boundaries overlap. Observation at this level is currently only possible through telemetry, where the chief deterrent is cost, particularly if several individuals are to be monitored simultaneously instead of sequentially. Ludwig (1984) reports that male *Microtus richardsoni* associate closely with groups of females, at least at high density.

Copulation Rate

Vehrencamp and Bradbury (1984) suggested that males could increase their copulation rate by:
(1) provoking mate choice by advertisement;
(2) by provisioning a female or otherwise enhancing her condition by assuming some of her costs;
(3) forcing copulation; and
(4) accelerating physiological receptivity by induction of abortion or infanticide in females whose young they are unlikely to have sired.
All these traits probably occur in microtine populations, and may be augmented by the universal induction of ovulation by males, and the capacity of males to accelerate puberty. Field studies of changes in sexual receptivity and development are almost completely lacking. By contrast, laboratory studies of a number of rodents have achieved a level of specificity and sophistication which suggest that a pheromonal role in endocrine regulation of reproduction is unarguable. While these results will be fully reviewed in Chapter 5, the following examples are illustrative. Carter *et al.* (1980) subjected virgin female *Microtus ochrogaster* to a variety of social conditions and stimuli for six days and then killed them to obtain uterine weights. The conditions involved housing the female (1) with a sister; (2) alone; (3) alone but with the urine of a castrated male applied to her nose once daily; (4) with a castrated male; (5) alone but with daily application of urine from an intact male; (6) with a castrated male but with daily application of urine from an intact male; (7) with an intact male for one day, and then alone; and (8) with an intact male for the whole period. The results are shown in Figure 3.1, and clearly indicate that male urine is

Table 3.1: Behavioural options for increasing fitness components of mating systems (modified from Vehrencamp and Bradbury 1984)

Components of fitness	Options for increasing fitness
Males	
(1) Female encounter rate	Stay near resources females need or pass through
	Remain with females once encountered
	Search for females using olfactory cues
(2) Copulation rate	Accelerate receptivity through abortion or infanticide
	Assume some costs of the female
	Provoke mate choice through advertising
	Force copulation
(3) Fertilisation	Exclude males by territoriality or dominance
	Displace another male's sperm
	Form a sperm plug
	Guard the female after insemination
(4) Female fecundity	Select mates with high fecundity
	Provision female
	Assume female's costs
(5) Juvenile survival	Select appropriate females
	Invest territoriality and/or paternal care
(6) Adult survival	Defer breeding to later ages
	Reduce investment in above to minimise risk
	Defend territory for resource access
	Form liaisons to dilute costs or risks
(7) Nepotistic term	Modify investment to increase relatives benefit
	Direct behaviour preferentially towards or away from kin
Females	
(1) Encounter rates	Options are similar to those for males
	Force males to aggregate by diferential choice
(2) Female fecundity (sperm are limiting)	Increase encounter rates
	Exclude other females by dominance or territoriality
(3) Female fecundity (resources are limiting)	Exclude other females
	Form liaisons with territorial males
(4) Juvenile survival	Maternal care
	Choose mate which is dominant, has a good territory, is a competent parent, or is genetically suitable
(5) Adult survival and nepotism	As for males
(7) Sexual selection term	Invest in mate selection using appropriate cues
(8) Demographic term	Modulate age of first breeding to maximise demographic payoffs in a fluctuating environment

Figure 3.1: The effect of the presence of males and their urine on uterine development in Microtus ochrogaster. Females were housed alone, or with female siblings, or exposed to a strange intact male, a strange castrated male (broken symbol), the urine of a strange male or the urine of an intact male in various combinations. After Carter et al. (1980)

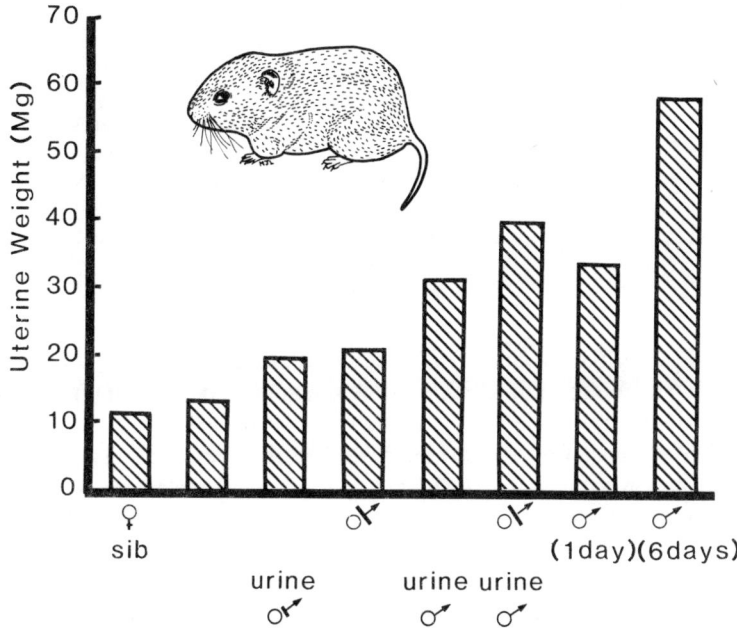

necessary for reproductive activation. The high values for condition (8) are probably a reflection of ovulation and pregnancy. Comparable effects are known from species with a variety of social systems (e.g. Baddaloo and Clulow 1981). *Microtus ochrogaster* males assist in territory defence and help females in a variety of ways (Thomas and Birney 1979; Getz and Carter 1980), but it is unclear whether this help accelerates return to oestrus in the female. Olfactory cues are implicated in sexual receptivity in many small microtines as a means of provoking choice (Hoffmeyer 1982; Huck and Banks 1982a, 1982b; Coopersmith and Banks 1983; Viitala and Hoffmeyer 1985). While there is evidence in some species of microtines that reproductive females can dominate males and exhibit choice (Banks, Mankovich and Huck 1979), it seems certain that the converse is also occasionally true. Under these circumstances forced copulation is likely, but impossible to establish in the wild given the current sophistication of field studies (Kawata 1985a).

Both infanticide and termination of pregnancy in the presence of a strange male are also well known from microtines and many other rodents, such as house mice. For example, male *Microtus agrestis* depress prolactin production in pregnant females and cause abortion (Milligan, Charlton and Versi 1979). In some microtines, this effect is very generalised (Schadler 1985). Because of the importance of these patterns, a brief aside is warranted to consider their recent interpretations.

Table 3.2: Five hypotheses for the development of infanticide. (Classification follows Hrdy and Hausfater 1984.)

Class of infanticide	Relatedness	Nature of gain
(1) Exploitation as a resource	Usually distant	Nutritional gain
(2) Competition for resources	Distant	Resources for self and kin
(3) Sexual selection	Distant	Increased breeding opportunity
(4) Parental manipulation	Close	Increased fitness for one or both parents
(5) Social pathology	Generally irrelevant	No direct benefits, but density may decrease

Infanticide.

The role of conflict in the evolution of mammalian social systems is excellently illustrated by the occurrence of infanticide. Although infanticide has been interpreted as social pathology induced in the laboratory by overcrowding (Calhoun 1962), there is now substantial evidence from observational studies of unperturbed populations of many mammals that infanticide may be both rather common, and amenable to adaptive interpretation (Hrdy 1979; Sherman 1981; Hausfater and Hrdy 1984). Hrdy and Hausfater (1984) provide the most useful classification of functional classes of infanticide (Table 3.2). They include within the ambit of the term of infanticide, action against gametes, embryos and young (the entire period of parental investment), and killing by both strangers and parents.

The most useful field studies come from observation of diurnal populations where genealogy of the individuals are known, which is rarely the case for microtines. In particular, studies of primates and ground squirrels have proved illuminating. Until recently, there has been some consensus that sexual selection, according to Hrdy's classification, is the most usual circumstance leading to infanticide. In addition, it has been argued that individuals are much more likely to kill unrelated young than related young. One well documented exception exists. Hoogland (1985) demonstrated that infanticide was not only exceptionally common among black-tailed prairie dogs, but generally directed against relatives. He concluded that the principal benefit of infanticide was exploitation of young as food, as infanticidal mothers maintained condition much better than non-infanticidal mothers during the breeding season. Although there must be some conflict of interest in killing relatives, Hoogland suggested that travel away from the local kin groups exposed mothers to the risk of losing her own young to marauders. Except in the case of exploitation for food, killing of kin is unlikely because killing relatives conflicts with the genetic interests of the adult.

Among the small rodents which are the focus of this study, laboratory analyses appear to confirm this pattern. Because this topic has been reviewed extensively (papers in Hausfater and Hrdy 1984; Labov, Huck, Elwood and Brooks 1985), and I agree with the thrust of these reviews, I will confine myself to a discussion of major generalities, prospects for future research and chinks in the armour of the current consensus. First, it does appear that relatives are usually less likely to be killed than non-relatives. In the laboratory, it has been shown on many occasions that males kill the offspring of females whose young they are unlikely to have sired. There appears to be some inhibition of infanticide associated with male familiarity with a female, though its causes remain controversial. It has been suggested that recent copulation or past sexual association is necessary for inhibition (e.g. Mallory and Brooks 1978; Labov 1980; Huck, Soltis and Coopersmith 1982), but other data support the notion that familiarity facilitates female dominance of the male (Elwood and Ostermeyer 1984; Wolff 1985; Elwood 1986). There is currently no consensus over whether testing a male in its home cage produces a different result to testing in a strange or 'neutral' arena (Beilharz 1975; Brooks and Schwarzkopf 1983).

Several studies have revealed differences in infanticidal tendency between laboratory strains of house mice, leading to speculation that this tendency may be under genetic control (reviewed by Labov et al. 1985). However, it may be that these effects are an artefact of endocrinological changes caused by inbreeding and artificial selection. Evidence that dominance exerts an overwhelming influence on infanticidal tendency in male mice supports an endocrinological cause (Huck et al. 1982).

At least for infanticide by adult males, there is widespread agreement that any benefit accrues predominantly to the male as a consequence of elimination of offspring he could not have sired, and acceleration of the return of the female to oestrus, enhancing his own mating opportunities. This is the sexual selection hypothesis of Hrdy and Hausfater (1984). Among the remaining classes of infanticide, there is currently no evidence from house mice or microtines that infanticide is used for food (exploitation hypothesis), or to enhance local competition for resources. In the latter case, enforced dispersal is more likely than the death of one of the contestants. The remaining class proposed by Hrdy and Hausfater (1984) is parental manipulation, with parents directing infanticide against one sex to ensure optimum sex allocation. Its importance remains controversial, and discussion will be deferred to the section on sex allocation in Chapter 6.

In summary, most of these ideas have their genesis in laboratory studies, and the significance of infanticide in field populations remains unquantified. There is good evidence that there is infanticide among field populations of microtines (Labov et al. 1985; Caley and Boutin 1985). However, there is equally good evidence that infanticide by males is of negligible demographic significance (Brooks 1984), as removal of males has little effect on juvenile recruitment (Boonstra 1977). Brooks (1984) attempts to salvage the notion that infanticide might be of significance in driving cycles by pointing to the importance of patterns of juvenile recruitment in influencing rodent demography, and suggesting that females may be the prime perpetrators of infanticide. The evidence for this hypothesis is negligible for rodents, but somewhat stronger for lagomorphs.

It is worth mentioning that several strands of mathematical theory now predict the evolution of infanticide. Because these analyses are rather preliminary, a full review will not be presented here (readers are referred to Hausfater 1984, 1986; Glass, Holt and Slade 1985; Glass 1986). One concept which appears worthy of further analysis is cohort selection, which Mertz, Craig, Wade and Boyer (1984) describe as frequency-dependent natural selection that takes place because of interactions among age groups within a population, where there is some genetic differentiation between the age groups. Mertz *et al.* (1984) contend that infanticide may occur as a consequence of cohort selection for 'spiteful' behaviour (*sensu* Hamilton 1970) with little or no benefit to the older infanticidal age class. Genetic differentiation may arise between age groups if the seasonal generations in microtine populations are under selection.

Bruce effect.

While infanticide may be of dubious demographic significance, its role in shaping the evolution of microtine social interactions should not be underestimated. It has been known for some time that exposure of female house mice to a strange male at critical stages of pregnancy (usually rather early), leads to the abortion or resorption of the embryos (Bruce 1959, 1960). The effect has subsequently been confirmed in many species of rodents, including many microtines (Clulow and Clarke 1968; Clulow and Langford 1971; Clarke and Clulow 1973; Stehn and Jannett 1981; Clulow, Franchetto and Langford 1982; Huck 1984). Preliminary analyses suggest that the Bruce effect is not present in lagomorphs (Bell and Reece 1983). Discussions of this phenomenon initially concentrated on male advantage (early return to oestrus in the female) (e.g. Trivers 1972; Wilson 1975; see however Bruce and Parrott 1960), or dismissed the result as an artefact of laboratory conditions (e.g. Bronson 1979). More recently, attention has been paid to possible advantages for the female. Schwagmeyer (1979) correctly points out that the evolution of male capacity to influence physiological processes in a female which may or may not be pregnant, without any defensive evolutionary response on the part of the female, is implausible, and also rather chauvinist (see also Wiley 1983; Krebs and Dawkins 1984).

Schwagmeyer (1979) and Labov (1980, 1981a) have identified a number of possible sources of benefit to the female:
(1) enhancing outcrossing (see Bruce and Parrott 1960), by maximising mating with strangers. She rejected this hypothesis because there are cases when the original male would be a better mate than the new one. However, Yamazaki *et al.* (1983) report the intriguing observation that the Bruce effect in inbred house mice was more pronounced when the stud and strange male differed in the H-2 region of the Major Histocompatability Complex (the MHC), than when the stud and strange male had the same H-2 type. The significance of these results remains unexplained, as strange females were also shown to induce pregnancy block when their H-2 type differed from that of the studs;
(2) as a way of acquiring a dominant mate. Evidence for this hypothesis is equivocal (Huck *et al.* 1982; Labov 1981b);
(3) as a prophylactic, or female counterstrategy, against infanticide. According to this view, it may be advantageous to terminate parental investment which has a low

prospect of success because of the risk of infanticide, particularly if this can be achieved early in pregnancy when the initial investment is slight, and when the risks to the mother are low. Huck (1984) provides a good review of the data which support this hypothesis, and proposes that it is tentatively worth accepting for house mice and most microtines;

(4) as a response to desertion, as a way of replacing a father unlikely to provide paternal care, with a male which might. Schwagmeyer (1979) attributes this hypothesis to Dawkins (1976), although I am uncertain whether the attribution is correct (as apparently is Dawkins 1981). The chief argument against this hypothesis is the very low importance of paternal care in most small rodents with polygynous or promiscuous mating systems (see however Barnett and Dickson 1985). It is certainly interesting that susceptibility to pregnancy block is much more prolonged in *Microtus ochrogaster* than in other microtines (Stehn and Richmond 1975; Kenney *et al*. 1977; Stehn and Jannett 1981). As this species is monogamous, and paternal care reasonably well developed, it may be the case that the Bruce effect common to most microtines has been elaborated in this species for the reasons proposed by Schwagmeyer (1979). These results also provide evidence, albeit tinged with the stigma of circularity, that monogamy in *M. ochrogaster* is obligate, rather than a facultative consequence of the juxtaposition of male and female territoriality.

Evidence for the existence of pregnancy block in the field is poor and largely inferential (Mallory and Clulow 1977). Although low fertility is commonly documented, it is very difficult to distinguish between the effect of pregnancy blocking

Figure 3.2: The frequency of overlap of home ranges of Microtus californicus *determined by telemetry. Overlap may vary between zero (no overlap) and two (complete coincidence of range). (Male-male = unshaded bars; female-female = dotted; male-female = slanted lines). After Ostfeld (1986)*

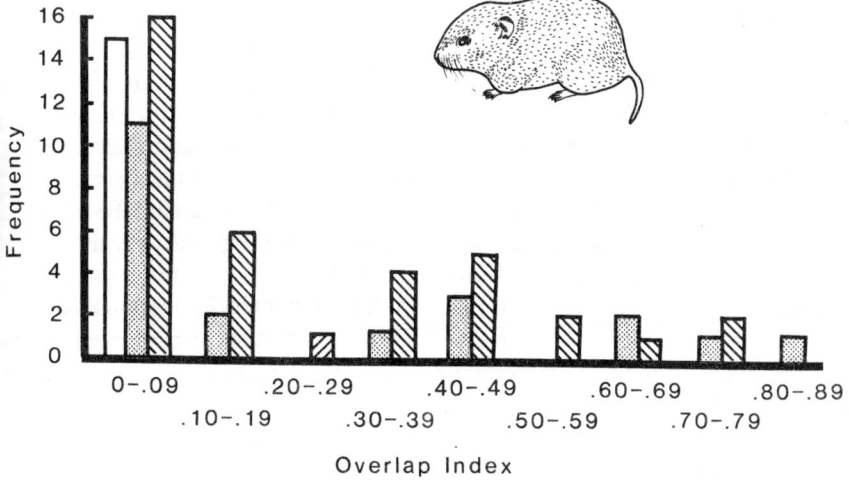

and matings which do not lead to pregnancy (Westlin and Nyholm 1982). At least in *Clethrionomys*, repeated mating over a number of sterile cycles serves to prime female reproduction, apparently by stimulating prolactin release (Westlin 1982a, 1982b; Westlin and Gustafsson 1983; Gustafsson 1985).

Heske and Nelson (1984) conducted an intriguing experiment in small (1.25 x 3 m) enclosures, in which *M. ochrogaster* females were able to reject or avoid a strange male, and where the original stud male was allowed to remain and defend his mate. They found evidence of pregnancy block in both conditions, and no evidence of infanticide under either case. Their data are contrary to both hypotheses (3) and (4) above, but do provide weak support for field significance for the Bruce effect. Clearly, the nature of pregnancy block requires further investigation, particularly with respect to interspecific differences, and the significance of the phenomenon in the field.

Fertilisation Rates

The easiest way to ensure paternity of the offspring of all females encountered is to maintain exclusive access to them. Some microtines exhibit rigid male territoriality where the male range completely encloses the range of one or more females. Use of trapping records to define the existence of territoriality is fraught with difficulty, as true home ranges and the ranges revealed by trapping are known to have limited coincidence (Stoddart 1970). As trapping almost inevitably underestimates range size, territoriality may be incorrectly inferred. Ostfeld (1986) was able to use radiotelemetry to investigate home range overlap in *Microtus californicus*, and demonstrated conclusively that while there is substantial male-female and female-female overlap in home ranges of this species, male-male overlap is virtually non-existent (Figure 3.2). Because ranges of these voles are very small (< 250 m^2), active territorial defence is implicated (see also Ostfeld 1985b).

Although these data are convincing, it is also clear that male home ranges overlap considerably in many other microtine species. This conclusion has also been well supported by radiotelemetry (e.g. *M. pennsylvanicus*, Madison 1978, 1980a, 1980b). While male territoriality ensures fertilisation of all females within the territory, in this latter case other methods of increasing fertilisation rate are implicated. One means of achieving this option would be dominance of nearby males. The definition and establishment of dominance has also proved problematic (Kaufmann 1983). If dominance is defined as 'a relationship between two individuals in which one (the subordinate) defers to the other (the dominant)...' Kaufmann (1983, p. 2), the problem of context to which I alluded in Chapter 1 needs attention. If the outcome of an encounter is dictated by familiarity or degree of association with local conditions, contrived encounters in a neutral arena may give misleading or meaningless results. However, some analyses are sufficiently consistent and plausible to suggest biological relevance for dominance in microtine social behaviour, particularly where home ranges of animals overlap or coincide. Shapiro and Dewsbury (1986) allowed pairs of conspecific male *Microtus montanus* to become familiar with separate small cages (29 x 19 x 13 cm) and then examined their use of a neutral area between the cages. Dominance was assigned when a male repeatedly attacked the other male, or where it both chased the other male at least twice and maintained exclusive use of the neutral arena for at least 70 per cent of the test. A sexually receptive female was then placed in a cage where she had the opportunity to associate with either the dominant or subordinate male. Similar tests were performed with *M. ochrogaster*. Females preferentially associated with the dominant in *M. ochrogaster* but not in *M. montanus* (Figure 3.3). Other similar tests support the generality of this difference between the two species (Shapiro and Dewsbury 1986). These results not only support the use of dominance in mate choice by females (see also Madison 1980b; Huck *et al.* 1981; Webster and Brooks 1981; Hoffmeyer 1982; Huck and Banks 1982a, 1982b; Viitala and Hoffmeyer 1985; Shapiro, Austin, Ward and Dewsbury 1986), but draw attention to the possibility that microtine social behaviour may show substantial interspecific diversity.

Determination of paternity and maternity.
The question of sperm competition is complicated, and can only be resolved if paternity is known. While there is good evidence of female promiscuity in

Figure 3.3: Preference by female Microtus montanus *and* M. ochrogaster *when free to associate with a dominant (stippled bar) or subordinate (slanted lines) male. After Shapiro and Dewsbury (1986)*

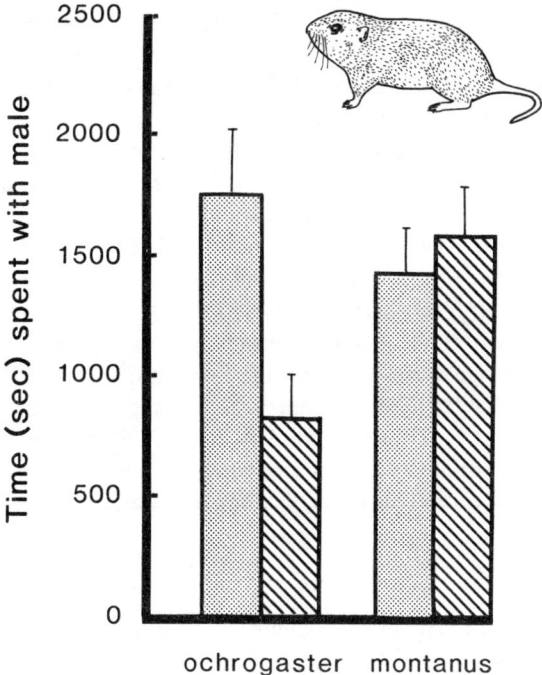

microtines (Kawata 1985a), there is no published evidence of multiple paternity. Indeed, techniques for the unambiguous determination of paternity in promiscuous species have only recently become available. The most commonly used technique relies on allelic variation in the structure of genetic loci coding for proteins and enzymes. Paternity determination depends on detecting sufficient variation to enable the combination of alleles found in an offspring to be contrasted with that of its mother to determine which alleles are paternally derived. If only one male from the pool of potential partners could have contributed these alleles, paternity is assumed. (For a very clear exposition of this technique see Ellstrand 1984.)

Three problems require mention. First, in most circumstances it is desirable to sample tissue for electrophoresis nondestructively. However, the number of enzymes available from blood or muscle tissue is limited, compared to tissues such as liver and kidney. Sheridan and Tamarin (1985) report the successful use of salivary amylases in paternity studies in *Microtus pennsylvanicus*, and Pavone and Boonstra (1984) report a technique involving surgical removal of a single kidney. Second, in field studies the mother may also be uncertain, if young are only captured after weaning (e.g. Cockburn, Scott and Scotts 1985). Tamarin, Sheridan and Levy (1983) have circumvented this problem by labelling the mother with a unique radionuclide which can be distinguished by whole body counting of offspring. While this technique is of considerable promise, it is not useful in remote sites, and can only deal with a small sample of females at any time. In species which nest communally, it may also infer maternity incorrectly (McShea and Madison 1985). Third, in species where males make occasional forays (e.g. Stoddart 1970), it may only be possible to assign probabilities to paternity (e.g. Foltz and Hoogland 1981; Kawata 1985a; see also Scott and Tan 1985). Identification of multiple paternity is a little easier. It is enough to know that two offspring in a litter have different sires, so the paternity of all the young need not be known to establish presence or absence of multiple paternity. Multiple paternity can also be proved by the detection of three different paternal alleles in the offspring of a mother. Although these techniques are only just being applied to microtine populations, there is some evidence that they are in the process of being rendered redundant. Advances in technology for manipulation and analysis of DNA have recently been applied to analysis of both paternity and maternity. Some elements of DNA are highly variable in both structure and extent of repetition and dispersion within the genome, and this variation greatly exceeds that detectable by protein electrophoresis. It is possible to construct an individual specific 'fingerprint' of these variable DNA fragments which then enables easy determination of parentage. Suitable variation is found in both satellite sequences within nuclear DNA and mitochondrial DNA (Kessler and Avise 1985; Wilson *et al.* 1985; Jeffreys *et al.* 1985). Mitochondrial DNA has haploid maternal inheritance and should prove useful for determining maternity. Jeffreys *et al.* (1985) have developed the use of nuclear satellite DNA to determine both maternity and paternity in human populations unambiguously. However, the technique does seem capable of application to other species, including rodents (R. Boonstra, P.A. Boag, pers. comms.). As my discussion of evolution of life histories and social behaviour proceeds, it will become clear that a knowledge of genealogy promises to advance our understanding of microtine biology profoundly, hopefully justifying this rather lengthy aside.

Electrophoretic determination of paternity in microtines has thus far focused on populations of species in which males, and not females, are territorial. The absence of evidence for multiple paternity suggests that it is much less important in microtines than in some other rodents, such as ground squirrels (e.g. Hanken and Sherman 1981). This does not mean that sperm competition is unimportant. The use of copulatory plugs (Hartung and Dewsbury 1978), and repeated copulation as a form of mate-guarding (Kawata 1985a), implies that there has been selection to prevent sperm displacement, particularly where the interests of the male and female are in conflict (see Dewsbury 1982b, for a review of the benefits and costs to the female of mating with several males). Milligan (1979) points out that the backwardly directed spines found on the penis of microtines may function to scrape out copulatory plugs from previous ejaculations. It is not clear whether the principal function would be to remove plugs produced by an earlier ejaculation within a single intromission, or to get rid of plugs produced by other males. Intraspecific diversity in the duration of copulations suggests that both effects may be important (e.g. Milligan 1975).

It is interesting to observe that the benefits of guarding are frequency dependent and a game theoretical approach may be appropriate for exploration of this strategy. Consider an animal which has the choice of seeking another mate, or staying and guarding a female through the period where sperm could be displaced or embryos aborted. The payoffs will depend on the encounter rate with other females and the proportion of successful embryos sired as a result of a copulation of varying length. The latter function will depend on the extent of interference by other males. Parker (1984) calls contests of this sort 'alternative-option scrambles', and had earlier presented a model of mate searching by male dungflies which has some analogies to the microtine case (Parker 1970, 1974, 1978). These observations reinforce a key conclusion of recent theory on mating options: a single solution appropriate to all circumstances may be an unrealistic expectation; instead, the appropriate behaviour may depend upon what other animals are doing at the time.

Female Fecundity

Where females of different age and size differ in fecundity, there may be strong proximate benefits to the male in preferring to mate with those females which are most fecund. Effects of this sort are unknown for microtines, but there is some evidence that males may have to choose between a sample of possible mates if female receptivity is synchronised. Kawata (1985a) examined a population of *Clethrionomys rufocanus* where the females were territorial, but males had ranges which overlapped those of other males and those of several females. Reproductive synchrony was pronounced and maintained as a consequence of postpartum oestrus. Using electrophoretic determination of paternity, Kawata concluded that an individual male was unable to fertilise two females within his home range when oestrus was synchronised. Male choice in these situations may warrant further investigation. The relation between age and fecundity is not entirely clear, although microtines probably conform to the mammalian pattern of a slight increase in litter size during the first few attempts, followed by reasonably stable litter sizes, and then a decline towards the end of their life (e.g. Leslie, Tener, Visozo and Chitty 1955; Negus and Pinter 1965; Hasler and Banks 1975). The significance of this late decline is uncertain, as laboratory animals are unusually long-lived and may be grossly unrepresentative of selective pressures likely to operate in the field.

Vehrencamp and Bradbury (1984) also suggested that a male might increase female fecundity by providing her with food or sharing some of her costs such as territorial defence. An effect of this sort seems most likely in monogamous species, but has not yet been reported from microtines.

Juvenile Survival

Investment by males may enhance the survival of offspring, just as it has the potential to increase female fecundity. In general, male investment by mammals is likely to be slight in comparison to the investment by females. This is because male mammals do not lactate, but the patterns are likely to be exaggerated by uncertainty about paternity in the polygynous and promiscuous mating systems exhibited by most mammals. Dewsbury (1985a) catalogued the types of care which could be exhibited by a male rodent. These include huddling, grooming, retrieving, play, tutoring, provison of food and manipulation in the nest. Indirect investment would include nest construction, territorial defence and warning about predators. Among microtines, paternal care appears to be best developed in the species which appear to be commonly monogamous (*Microtus ochrogaster* and *M. pinetorum*, e.g. Thomas and Birney 1979; Gruder-Adams and Getz 1985; McGuire and Novak 1985; Oliveras and Novak 1986) and does appear to increase juvenile survival (Gavish, Carter and Getz 1981).

While these results are consistent with the prediction that paternal care will be most prevalent in monogamous mating systems, a surprising result was obtained from a study of cold adaptation in captive house mice by Barnett and Dickson (1985). They developed two strains of mice: (1) controls which were housed at 23° C; and (2) 'Eskimo' mice which were housed at 3° C. The fecundity and juvenile survival of the Eskimo strain was initially very low, but after ten generations the number of young weaned by an Eskimo pair approached that of the controls. To produce the result of interest, Barnett and Dickson contrived matings in the cold environment between each of the possible combinations of Eskimo and control parents. As anticipated, the control male-control female cross produced fewer young than other combinations. The surprising result came from analysis of survival to weaning. Both control females and Eskimo females mated to an Eskimo male reared a greater proportion of young than did females mated with a control male (Table 3.3). This implies that males contribute more to enhanced survival than females. As house mice normally mate promiscuously or polygynously, the assumption that male contribution is confined to monogamous sytems may be unwarranted.

Table 3.3: The mean number of young born and weaned per pair of control or Eskimo house mice in a cold environment (3°C) (data from Barnett and Dickson 1985). The influence of male type on weaning success was highly significant ($p < 0.003$), while female type had no influence ($p < 0.2$)

Male parent		Female parent	
		Control	Eskimo
Control	Born	17.7	32.4
	% Weaned	51	56
Eskimo	Born	28.5	28.0
	% Weaned	73	84

Adult Survival

If any of these options are costly for adult survival and hence reduce *future* mating options, it may be advantageous to defer breeding or reduce investment. Extreme intraspecific variation in the age at which males sexually mature is well known for microtines (for a comprehensive review see Gipps 1985), but its causes and adaptive significance remain extremely confused. Discussion of this topic will be deferred to Chapter 5.

Nepotism

One of the chief stimuli to the emergence of behavioural ecology was Hamilton's (1964) elaboration of Fisher's (1930) observation that the fitness consequences of many behavioural interactions may be influenced by the degree of relatedness between the interacting animals (see Chapter 1). Consequently, any of the interactions and options listed above may be altered by the degree of relatedness between competing or interacting animals.

Kin recognition.

In order to consider this hypothesis, a brief aside is necessary. The theory of kin selection depends on the assumption that animals are capable of distinguishing their relatives from non-relatives, or live in a society where this distinction is never necessary. As dispersal is a conspicuous component of the population biology of microtines, kin recognition should be important. Holmes and Sherman (1982, 1983) have analysed the criteria which could be used to distinguish kin and the way such criteria could develop. There are four mechanisms :

Recognition based on spatial distribution. For example, the majority of birds exhibit female dispersal and male philopatry (Greenwood 1980). If inbreeding avoidance is desirable, females should refuse to mate with males close to their birth site, as these males are likely to be relatives. Although a similar argument might be advanced for mammals, where males are the predominant dispersers (Greenwood

1980), available evidence is consistent with female manipulation of recruitment to minimise inbreeding, rather than male choice on the basis of kin recognition (Cockburn, Scott and Scotts 1985). It is certainly true that in many microtines, females exert greater control over recruitment than males (Bujalska 1973, 1985a; Boonstra 1977; Viitala 1977; Redfield, Taitt and Krebs 1978a, 1978b; Madison 1980b; Boonstra and Rodd 1983; Gipps *et al.* 1985). However, the generality of this conclusion is weakened by the focus on species in which females are territorial, and the appropriate experiments on species where males or both sexes exhibit territoriality need to be attempted (see below).

Recognition based on familiarity or association. If an animal is reared in a den, a nest or a burrow, it may be possible to use the probability that all burrow inhabitants are relatives as a criterion for kin identification. Holmes and Sherman (1983) feel that this is the most common basis for recognition, and this is almost certainly true for microtines. For example, in some species an incest taboo appears to reduce the probability of brother-sister matings (Table 3.4). *Microtus canicaudus* and *M. ochrogaster* individuals which are familiar with each other rear fewer litters than unfamiliar pairs, regardless of kinship (Gavish, Hofmann and Getz 1984; Boyd and Blaustein 1985; Fig 3.4). This is a clear example of kin recognition based on prior association. Non-random patterns of association observed in winter groups of *Microtus xanthognathus* are also indicative of preferential avoidance of relatives (Wolff and Lidicker 1981), but the cues used to identify relatives are less clear.

Holmes and Sherman (1983) point out that the capacity for discrimination in either of these two systems is limited. For example, might it be possible for a male to distinguish his own young among a litter of pups? This may be a problem in the

Table 3.4: Observations suggesting the presence or absence of incest avoidance between siblings in microtine rodents

	Mechanism	Source
Avoidance present		
Microtus ochrogaster	Transfer of urine necessary for estrus. Absent between siblings	Carter *et al.* (1980) McGuire and Getz (1981) Gavish, Carter and Getz (1983) Gavish, Hofmann and Getz (1984)
Microtus canicaudus	Low conception rate	Boyd and Blaustein (1985)
Microtus pinetorum	Low conception rate	Schadler (1983)
Microtus californicus	Not given	Batzli, Getz and Hurley (1977)
Lemmus sibiricus	Not given	Facemire and Batzli (1983)
Avoidance absent		
M. pennsylvanicus	None	Batzli, Getz and Hurley (1977) Porter and Dueser (1986)
M. oeconomus	None	Facemire and Batzli (1983)

Figure 3.4: The role of familiarity in incest avoidance in Microtus canicaudus (= M. montanus canicaudus). *Whether young are reared together or reared apart has a much greater effect on fertility than whether they are siblings (unshaded bars) or unrelated (slanted lines). Data from Boyd and Blaustein (1984)*

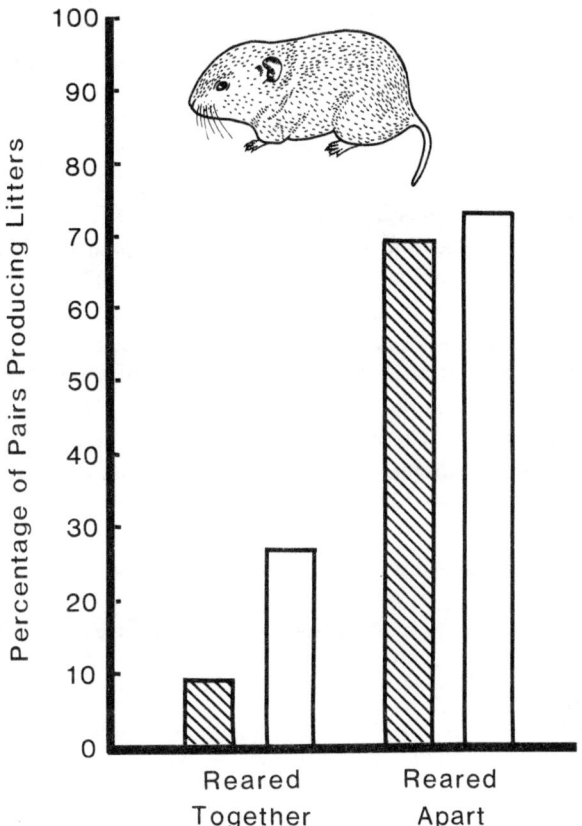

case of sperm competition and multiple paternity. Indeed it has been suggested that infanticidal tendencies in males may be suppressed if they have copulated with a female (e.g. Huck *et al.* 1982), though this remains controversial (Elwood and Ostermeyer 1984). Among microtines, the introduction of a strange male reduces female weaning success but introduction of the sire does not (Mallory and Brooks 1978; Webster, Gartshore and Brooks 1981).

It may be useful for true siblings to distinguish themselves from half-siblings. Hanken and Sherman (1981) showed that an extraordinarily large proportion (>76 per cent) of litters of Belding's ground squirrels (*Spermophilus beldingi*) were sired by more than one male. Yearling full sisters were less likely than distaff and spear half-sisters to fight and chase each other (Holmes and Sherman 1982; Holmes 1986).

The remaining two hypotheses have the potential to account for this complexity.
Recognition alleles. Hamilton (1964) originally proposed that recognition may be under allelic control. Dawkins (1976, 1982) has called this hypothesis the green-beard effect. If an animal has an extremely rare heritable trait e.g. a green beard, and can recognise that trait *both in itself and its relatives*, then it should be capable of distinguishing close relatives. The green-beard hypothesis has been ridiculed because of the plethora of functions the allele would need to serve (phenotypic expression of the trait, recognition of the trait in self, recognition of the trait in relatives), and the number of such alleles which would need to be present in a large panmictic population (Lewin 1984). However, cautious support for a modified green-beard effect is emerging (e.g. Blaustein 1983; Blaustein and O'Hara 1985). Indeed, what is currently the best evidence for the effect comes from an intriguing analysis of mate preference in the house mouse, and so is worthy of attention here. In a series of experiments by Kunio Yamazaki and his colleagues (reviewed by Beauchamp, Yamazaki and Boyse 1985), it has been demonstrated that male and female mice prefer to nest and mate with animals which differ from themselves in the H-2 locus. The discrimination is achieved through olfaction. This bias could operate in nature to promote heterozygosity at the H-2 locus. The MHC contains gene clusters responsible for immunological functions, including the H-2 and the T region. These gene clusters exhibit both extreme variability and mutability (e.g. Bodmer, Trowsdale, Young and Bodmer 1986). This variability has a number of biologically important consequences and has been implicated in hypotheses which attempt to explain the evolution and maintenance of sexual reproduction in terms of the capacity to resist disease (Hamilton 1982). Heterozygosity may be crucial in a species where effective population size is often reduced drastically by the seasonal effects alluded to in Chapter 2 (see also May 1986).

In the current context, one of the components of the variation in the MHC is a series of genes known as t-alleles. Most populations of mice contain individuals which are either homozygous for a wild-type allele (+/+), or heterozygous for a variable recessive allele (+/t). Most t-alleles are homozygous lethals, but even dissimilar alleles may cause sterility. The maintenance of these alleles, which may approach frequencies of 0.5, is a considerable problem (Lacy 1978). One factor promoting the high level of these alleles is segregation distortion, with males often transmitting their t allele to 90 per cent or more of their progeny. Mating with a heterozygous male is problematic for a heterozygous female because many of her offspring will be sterile or inviable. The ability to discriminate male genotype should be under strong selection. Lenington (1983) and Lenington and Egid (1985) confirm that such discrimination is possible in the female. However, subsequent analyses using recombinant mice have confirmed that males can only distinguish individuals at the H-2 locus and are incapable of discriminating the t-locus genotype of the female (Egid and Lenington 1985). Because the t-locus and H-2 locus are closely linked, the necessity for discrimination may not have evolved, or the criteria used by males in mate choice may be less pronounced.

The observation that rodents can use olfactory cues to distinguish the genotype of conspecifics, and distinguish that genotype from their own, supports the basic premise of the green-beard hypothesis, and suggests that extension of this research beyond the well known house mouse will be of great interest.

Phenotypic matching. Under this mechanism, an individual learns the characteristics of either itself or its close relatives. When first encountering a strange individual, it matches the individual against the learned templet, and directs affiliative or discriminative behaviour accordingly. The criteria necessary to demonstrate phenotypic matching are clear. First, animals should recognise kin with which they have not been associated, yet base this generalisation on learned aspects of their own phenotype. Holmes and Sherman (1982, 1983) argue convincingly that the ability to distinguish between full siblings and half-siblings in ground squirrels is an example of phenotypic matching. Greenberg (1979) describes an ingenious experimental programme which supports a similar conclusion about sweat bees. No relevant data are available for microtines, but this should not preclude further research on this mechanism.

Laboratory data support the idea that microtines recognise their kin, but current studies suggest that kin recognition often functions to promote avoidance rather than affiliation. Field data are totally inadequate. Any relevant research will be of value.

Options Available to Females for Increasing Fitness

Encounter Rates

Whether encounter rates ever limit reproductive success in female microtines is uncertain, and not worth pursuing here.

Female Fecundity

Although limitations to sperm are unlikely to be as important as limitations to ova, ovulation in female microtines is induced by vaginal stimulation from copulation or other similar stimuli, suggesting coadaptation in the reproductive requirements of males and females (Gray and Dewsbury 1973, 1975). Several ejaculations may be necessary for pregnancy in *Microtus* (Kenney, Hartung and Dewsbury 1979). Although the number of ejaculations within a copulation in *Clethrionomys* appears to be low, and dictated by female receptivity (Milligan 1979), several sterile cycles accompanied by non-fertile matings also appear to occur in pubertal females, possibly functioning to prime the female reproductive tract (Gustafsson 1985; Ims 1985). Given that the number of ejaculates that a male may produce is limited (Dewsbury 1982a; Dewsbury and Sawrey 1984), there may potentially be some competition for access to sperm, but the influence of this competition on female behaviour remains completely unexplored. Certainly the significance of repeated copulation needs to be analysed both from a male and a female perspective (see also Doty 1974).

By contrast, female defence of food and breeding resources through territoriality is well established in some species through telemetry (Madison 1980a), the use of radioisotopes (Stoddart 1970) and trapping studies (Mazurkiewicz 1971). Dominance relationships among female microtines have attracted less attention than relationships among males, and warrant more attention in species where females are not territorial, and which form extended matrilineal groups at high density.

Juvenile Survival

Maternal care in mammals is, of course, universal. The options for manipulation of that care are enormous, and will be discussed throughout the remainder of this text. Mate choice by females may also provide short-term benefits to offspring which enhance survival. (This will be discussed under *Sexual selection* below.)

Adult Survival

Like males, there should be conflicts between the advantage of maximising current reproduction and the future reproductive success of the female. Two points require mention. First, there should be conflict between the mother and her offspring over the amount of care she provides (Trivers 1974). It may even be advantageous for the mother to withdraw investment altogether. The widespread occurrence of abortion in response to the presence of a strange male could be regarded as adaptive from the females viewpoint, and may act as a prophylactic against subsequent infanticide (see above).

The best evidence that social liaisons may act to improve adult survival is the collapse of territoriality during winter in many populations, when females (and males) form huddling aggregations for thermoregulatory benefit (Madison 1984; West and Dublin 1984). Because these aggregations may persist into the breeding season (McShea and Madison 1984), cooperation among females to maximise survival needs to be considered. Such cooperation may be an integral part of the breeding biology of some species. Boyce and Boyce (1985, and pers. comms) have used telemetry to show that, at least when density is low, some *Microtus arvalis* females nest and breed in groups. About three females usually aggregate together, and all give birth synchronously, all donating a couple of young to a communal litter. The population also contains solitary females, but these females survive population crashes poorly. This indicates that aggregation acts to improve female survival, perhaps because of energetic savings from huddling, even if some reproductive costs (e.g. reduced litter size), are sustained.

Nepotism

The evidence pertaining to a nepotistic effect in microtines is currently identical for males and females.

Sexual Selection

Darwin (1871) was the first to recognise that variance in the number of successful matings created opportunity for selection to operate on members of one sex, and hence could contribute to sexual dimorphism. He recognised two distinct mechanisms through which such sexual selection could occur; intrasexual selection would lead to characteristics which enabled one sex to win contests over mates (e.g. weaponry), whereas intersexual selection would cause the evolution of characteristics which would enable individuals to attract mates. Both systems undoubtedly operate in nature, but distinguishing between the two is often difficult (see Halliday 1983 for a thoughtful review). While the evolutionary significance of intrasexual selection is uncontroversial, the evidence for mate choice driving evolution of characteristics of the chosen sex is problematic. For example, mate choice may be based on immediate benefits to the female, and these benefits need not be correlated with the genetic characteristics of the male, e.g. the male occupies a better territory in which to raise young. There has consequently been intense theoretical speculation and empirical analysis of the likelihood that at least some mate choice is based on the genetic characteristics of the partner, and that such choice enhances the 'genetic quality' of progeny. In analysing sexual selection, it is important to recognise that male traits are more likely to be affected than female traits. Male gametes are less likely to be in short supply than female gametes (Bateman 1948; Trivers 1972, 1985), though perhaps not to the extent that was originally believed (Dewsbury 1982b, 1985b; Nakatsura and Kramer 1982; Dewsbury and Sawrey 1984). This will be particularly true in the polygynous mating systems characteristic of mammals, where variance in male lifetime reproductive success is likely to exceed that of females greatly (Clutton-Brock 1983). Hence it is expected that females will exhibit selection for male traits. Choice for good male genes may have profound consequences for evolution (e.g. Fisher 1930; Lande 1981; Seger 1985; Seger and Trivers 1985). However, the central theoretical problem is how evolution will establish female preference for a trait which is indicative of good genes, particularly as population genetical theory suggests that variance in lifetime reproductive success will disappear before female preference is established (Hamilton and Zuk 1982; Maynard Smith 1985a). Unfortunately for this theory, the empirical evidence for choice is overwhelming (e.g. Bradbury, Vehrencamp and Gibson 1985), and needs to be considered in any analysis of female options for increase in mating success.

Among microtines, the evidence for female choice of males is increasing. I have already discussed evidence that some females prefer to mate with dominant males. This may be unrelated to selection of 'good genes', as a dominant male may be more likely to afford protection to the female from other males (but see Manning 1985 for an alternative view). Further, there is some evidence that dominant males are most likely to kill young they are unlikely to have sired, at least in house mice (Labov 1980). A female may save considerable time and effort by accepting as a mate a male which is dominant and therefore likely to be infanticidal.

Even the preference exhibited by female house mice for males with appropriate t-alleles may have evolved in response to a short-term benefit. Inappropriate choice could lead to homozygous t-lethals and lead to inviable or sterile offspring. A related form of mate choice is selection which functions to reduce the probability of incest. Some microtines do show evidence of an incest avoidance (Table 3.4), and the importance of this selection will be elaborated in Chapters 5 and 7.

Demographic Adjustments

Although males are unlikely to derive major gains by deferring maturity, Vehrencamp and Bradbury (1984) point out that females in unstable populations may be able to maximise demographic benefits by modulating the age of first breeding. For example, it is known that in expanding populations it is generally preferable to minimise the time taken to puberty, but this conclusion need not hold in declining populations (Mertz 1971). Modulation of the age of first reproduction by female microtines is known to occur in many circumstances. For example, very few females reach puberty in the season of their birth in declining populations (Myllymäki 1977b). The significance of this variation will be elaborated elsewhere.

Summary of Options for Increasing Fitness through Mating Systems

This brief review should support my contention that the study of mating systems in microtine species is of profound interest and complexity, and that some of the formidable problems of method are now being overcome. Although no study has ever measured more than a small fraction of these parameters of interest simultaneously, and while the data base is grossly deficient for most species, it is possible to conclude that mating and reproduction are characterised by substantial interspecific diversity. This variability will be addressed in the next four chapters. In Chapter 4, I attempt a conventional classification of mating systems, and seek to identify correlations between mating system, reproduction and habitat characteristics. In Chapter 5, I examine the factors which influence the onset of reproduction, contrasting hypotheses which implicate social suppression of reproduction with those which interpret delayed maturity from an adaptive perspective. In Chapter 6, I examine the disposition of reproductive effort. In Chapter 7, I look at the options that parents have for variation in reproductive investment through dispersal of their offspring, and how these may be integrated into strategic models of life history evolution.

Chapter 4
Territoriality, Space Use and Mating Systems

Territoriality is the more or less exclusive defence of an area by an individual or group (Davies and Houston 1984). Its demographic significance is clear, but often complex (Krebs 1971). Territorial defence is anticipated whenever defence of a spatially localised resource is less costly than ignoring other animals in the population and concentrating on efficient exploitation of that resource (Brown 1964 and many others). Resource defence may satisfy both short- and long-term objectives, and the latter may be very important in fluctuating environments (see Lima 1984 for a review). Interpretation of territoriality also requires recognition that the resource defended by the two sexes may be quite distinct. Trivers (1972) was the first to suggest that while female behaviour should be directed towards selection of a place appropriate for the raising of offspring, male behaviour should be directed towards the acquisition of mates. Further, while most early investigations of social behaviour have tended to seek species-typical behaviour, detailed investigations of many species of birds and mammals have led to increasing recognition that most species are characterised by diversity of territorial and mating behavior, both at an interpopulation and intrapopulation level. Perhaps the most instructive example of intraspecific diversity comes from a study of a small passerine bird, the dunnock, *Prunella modularis* . In a population of only 90 breeding birds in the Cambridge University Botanic Garden, Davies and Lundberg (1984) demonstrated that mating combinations included polyandry, monogamy, polygynandry and polygyny. In the order described, these mating systems represent increased male access to mates. Male control of access to mates in the dunnock depended on home range size, which was in turn influenced by the availability of dense patches of food. When food was dense and home ranges small, females were easily monopolised, and polygyny and polygynandry were prevalent. Where food was sparse, female ranges were large, and often two males shared a territory. In such circumstances, although one male was usually dominant to the other, the female would actively solicit copulations from the subordinate male (Davies 1985). Subordinates only helped the female when they procured copulations, and would even kill chicks or destroy eggs if denied the opportunity to mate. When both males helped, clutch size was increased, and nestlings got more food than did young raised by pairs.

These observations are relevant to the study of microtine territoriality in a number of ways. First, they support the notion of female territoriality being influenced by food abundance, and male territoriality by the dispersion of females. However, they clearly draw attention to conflicts which may arise between the sexes, and the way in which members of one sex may attempt to exploit the other. Some aspects of such conflict will be discussed in the section on infanticide, but further investigation of their role in microtine social systems will be invaluable.

Second, they clearly illustrate the importance of intraspecific diversity in interpretation of social systems. While it would be premature to suggest that comparable diversity is likely to occur in microtine populations, it may be over-reductionist to imply that particular populations or species are all exhibiting similar patterns of resource exploitation and mating tactics.

In the majority of microtine species, only one sex exhibits strong territoriality, or the exclusion of same-sexed conspecifics. In *Clethrionomys*, female territoriality appears to be universal (Table 4.1). However, in *Microtus*, there is considerable diversity in the sex bias and extent of territoriality (Table 4.1). Unfortunately there remain major deficiencies in our knowledge. For example in *M.townsendii*, the species whose demography has been analysed with the most elegant experiments (see e.g. Krebs 1979; Taitt 1985a), it is still not possible to say with confidence which territorial system prevails. The pattern of territoriality in any of the lemming genera remains unclear, as field studies have not been analysed appropriately (e.g. Banks, Brooks and Schnell 1975), and laboratory studies provide conflicting or inconclusive results (e.g. Bowen and Brooks 1978; Banks, Mankovich and Huck 1979; Semb-Johansson, Wiger and Engh 1979; Facemire and Batzli 1983).

The most comprehensive analysis of these patterns has been provided by Ostfeld (1985a). He suggests that there is a correlation between the nature and renewability of food resources and the sex which exhibits territoriality. Omnivorous species and animals which consume forbs generally show female territoriality, while species consuming fast-growing renewable foods such as grasses and horsetails (*Equisetum* spp.) exhibit male territoriality. Under some circumstances population fluctuations may alter the intensity of this pattern, but the extent of density-dependence is unknown.

Ostfeld (1985a) predicted that the cost of permitting intruders would be lowest when resources were abundant and quickly renewable. Grasses are likely to fall into this category, because they have numerous reproductive and somatic

Table 4.1: Patterns of territoriality during the breeding season in the microtines. Assignment is based on the most common pattern exhibited, and the existence of exceptions denoted. Modified from Ostfeld (1985a) and Viitala and Hoffmeyer (1985). (*A* = *Arvicola*; *C* = *Clethrionomys*; *L* = *Lemniscus*; *M* = *Microtus*)

Species	Technique	Variation	Source
Male territoriality (absent in females)			
M. agrestis	Trapping	No	Myllymäki (1977b)
M. californicus	Telemetry	No	Ostfeld (1986)
M. oeconomus	Trapping	?	Tast (1966); Viitala (1980)
M. xanthognathus	Trapping	?	Wolff and Lidicker (1980)
Male and female territoriality (boundaries coincident)			
M. ochrogaster	Telemetry	Yes	Hofmann, Getz and Gavish (1984)
			Getz and Hofmann (1986)
M. pinetorum	Telemetry	No	FitzGerald and Madison (1983)
Male and female territoriality (boundaries and nests not coincident)			
M. montanus	Telemetry	No	Jannett (1978, 1980, 1981, 1982)
Female territoriality (absent in males)			
A. terrestris	Isotopes	No	Stoddart (1970)
M. pennsylanicus	Telemetry	Yes	Madison (1980a,b)
			Madison *et al.* (1984)
M. richardsoni	Telemetry	Yes	Ludwig (1984)
C. gapperi	Trapping	No	Bondrup-Nielsen (1986)
C. glareolus	Trapping	No	Mazurkiewicz (1971)
			Bujalska (1973, 1985a)
C. rufocanus	Trapping	No	Kawata (1985a)
			Viitala (1977)
C. rutilus	Trapping	Yes	Viitala and Hoffmeyer (1985)
			Burns (1981)
Territoriality questionable (some evidence of communality)			
M. arvalis	Telemetry	Yes	Boyce and Boyce (1985)
			Mackin-Rogalska (1979)
L. curtatus	Trapping	?	Maser (1974)

adaptations which means that they are unlikely to be killed by grazing. Indeed, some authors have contended that grazing may promote compensatory growth which facilitates competition between clones of grass and the forbs with which they presumably compete for space (see Chapter 2). Because females of grass-eating species do not need to defend a territory in order to ensure that resources are renewed, their distributions may overlap. By contrast, plants which are killed by browsing, or food resources which are replenished only once each year are likely to have very low rates of renewability, and in some circumstances regrowth is unlikely in the lifetime of the small grazing animal. In these circumstances, the cost of resource depression may outweigh the cost of defending a territory against intruders, and female territoriality should evolve to ensure adequate food supplies for the rearing of offspring.

The theoretical relations between resource renewal and territoriality are proving to be quite complex, and to depend on both the currency of optimisation (maximising or minimising), whether feeding time or food processing is constrained, and on the shape of the function relating cost to territory area (e.g. Waser 1981; Brown 1982; Schoener 1983). Indeed, some reviews have concluded that renewal of food or prey *promotes* territory defence (e.g. Alcock 1984). Empirically, these relations remain better understood for insectivores and nectarivores than for herbivores (Pyke 1979; Waser 1981; Davies and Houston 1981, 1983; Houston, McCleery and Davies 1985; but see Hölldobler and Lumsden 1980). If Ostfeld's contentions are correct, microtines may prove to be useful models for understanding the evolution of territory size and defence in small herbivores. Processing constraints may be of particular importance in these taxa, because of the difficulties in digesting plant tissue (Chapter 2).

Unlike females, male reproductive success in mammals is more likely to be influenced by access to females than by access to food. As most microtines contribute little paternal care, Ostfeld (1985a) argues that male fitness will be limited by access to reproductive females in oestrus. He further suggests that renewability of oestrous females is likely to be constant across microtine species, but that the dispersion of females is not (e.g. Cockburn and Lidicker 1983). Where females are clumped in response to clumping of food, it may be possible for males to defend these clumps. Regular dispersion is probable among territorial females, and their defence is therefore unlikely. By contrast, females whose ranges overlap greatly can have much more contagious distributions, facilitating the development of male territoriality.

A problem with assessing the strength of comparative analyses of this sort lies with the choice of taxonomic unit which is used as a replicate. Ridley (1983) has suggested that a useful means to analyse hypotheses of correlation between a behavioural trait and environmental circumstances is to estimate the number of times that trait has arisen in response to those pressures (see also Felsenstein 1985; Cheverud *et al.* 1985). Hence, the uniformity of female territoriality in *Clethrionomys* can only be used as a single case in support of the idea that females defend resources which are renewed only slowly. The contrary data provided by the grass-eating *Arvicola terrestris* would be assigned equal weight according to these principles, leaving the isolated cases of *Microtus pennsylvanicus* and *M. richardsoni* the only remaining data in support of the hypothesis. However, if anything, interspecific and intraspecific variation in the diet of *Clethrionomys* exceeds that of *Microtus* (compare Batzli 1985 with Hansson 1985a, 1985b), rendering intergeneric analyses of questionable significance.

Ostfeld's hypothesis does appear to be a reasonable descriptor of interspecific diversity within the genus *Microtus*, and he does suggest a number of useful extrapolations of his model which appear amenable to direct testing. These include the idea that a 'threshold' for female territoriality should exist in all species, but will vary with the nature of the diet. He also suggests that as males are capable of devoting a greater proportion of their energy budgets to territorial defence than are females, males should be influenced less by intruder density.

While agreeing that there are likely to be strong differences in the factors which limit fitness in males and females, and that food dispersion is likely to underly these patterns, I believe that the synthesis could be enhanced by giving attention to temporal variation in mate availability as well as food availability. Ostfeld characterises the availability of reproductive females as temporally constant between species, but spatially variable. I believe this to be an oversimplification of both interpopulation and interspecific differences, and to ignore the substantial seasonal fluctuations in the availability of sexually receptive females. In particular, where the onset of reproduction is synchronised because of the seasonal fluctuations characteristic of the northern latitudes where microtines dwell, there is likely to be a peak in sexual reproduction not repeated at other times of year. Synchrony may affect mating opportunities in a variety of ways. Kawata (1985a) showed that where two territorial females within the home range of a male red-backed vole (*Clethrionomys rufocanus bedfordiae*), synchronised parturition and postpartum oestrus, the male was unable to inseminate both females. While experience may allow some older females to achieve an early oestrus because their reproductive tracts are better developed or 'primed' (see for example, Cockburn 1981), this first burst of reproduction is unlikely to be repeated. Subsequent reproduction and oestrus in older females is likely to be affected by metabolic debilitation associated with the costs of reproduction, and the extent to which intraspecific interference leads to early termination of some attempts at rearing. Reproduction in primiparous females breeding later in the season will be greatly affected by social interference from conspecifics, and most of these effects will break down synchrony (see Chapter 4).

Classification of Social Organisation Among Breeding Microtines

It is clear from the preceding discussion that several classes of mating systems are prevalent among the microtines, and that in part these are a reflection of the great differences in patterns of territoriality within the sexes. It is appropriate to consider to what extent these differences are associated with other unique aspects of the reproduction, habitat and social behaviour of the species in question. Unfortunately the data needed to assign correlations of this sort are fragmentary and often inappropriate to answer this question. In particular, I would draw attention to the recent spate of publications which contrasts *M. ochrogaster,* for which the evidence for monogamy is now overwhelming, with another *Microtus* species, apparently based on the assumption that the polygynous or promiscuous behaviour of these species are influenced by the same factors (e.g. Wilson 1982a, 1982b). This is clearly untrue, as polygynous systems also show great variation. Another unfortunate historical accident has been the early sophistication achieved in the study of *M. pennsylvanicus* (Webster and Brooks 1981; Madison 1980a, 1980b), which is apparently unusual within the genus *Microtus* (Table 4.1). Coupled with the elegance of European studies on *Clethrionomys* biology, an overemphasis on female territoriality has frequently appeared in the literature (e.g. Brooks 1984).

Following Emlen and Oring (1977), it is possible to recognise at least five distinct social systems within the microtines (Table 4.2). Note that I follow the Emlen and Oring criteria uncritically. In particular, I have not used the categorisation by Ostfeld (1986, 1987) which calls the combination of male territoriality and female home range overlap mate-defence polygyny. I do accept some of the caveats he suggests about uncritical assignment within these

Table 4.2: The relation between patterns of territoriality in the microtines and the classification of mating systems proposed by Emlen and Oring (1977)

Mating system	Territoriality		Relation of male range to female range	Males per group or territory
	Male	Female		
Polygyny				
(a) Resource-defence				
Female territories	Yes	Yes	Greater	One
No female territories	Yes	No	Greater	One
(b) Male-dominance	No	Yes	Greater	One
Monogamy	Yes	Yes	Equal	One
Polyandry	Yes	Yes	Equal	> one
Communality	Correlates not well documented			> one

categories (Ostfeld 1987), but feel the use of my classification is preferable until more precise experiments resolve the difficulties in interpreting male behaviour. *Microtus pinetorum* may either represent a sixth system (polyandry), or is monogamous (FitzGerald and Madison 1983). As polyandry cannot be disproved with available data, and may have profoundly different evolutionary implications, I will assume for the sake of argument that such a system does occur within the microtines.

Before attempting to classify species according to the mating system they exhibit it is worth commenting that intraspecific variability is rife. Unsurprisingly, density dependent effects have attracted most attention. For example, Ludwig (1984) suggested that at low densities male *Microtus richardsoni* moved freely between females, but at high densities were normally associated with a small clump of females. In several species, otherwise polygynous animals approach a monogamous condition at very low densities, presumably because mates are in such short supply that polygyny is impossible (e.g. *M. californicus* - Lidicker 1979, 1980; *M. montanus*- Jannett 1984). Despite this variability, it is possible to assume that each species except *M. arvalis* and *Lemniscus curtatus* (see e.g. Mullican and Keller 1986) has a modal mating system, and I am unaware of any complete breakdown of mating system at high density in an unconfined population.

Correlates of Variation

Variation within microtine social systems probably affects males in two important ways:
(1) through variation in the extent of access to mates; and
(2) through variation in male certainty about paternity.
Each of the six recognisable social systems may be assigned along these axes (Figure 4.1). For example, while a monogamous male probably has high assurance of paternity, he also has restricted access to mates. Males which are territorial and defend resources which females require will have greater assurance of paternity than non-territorial males, where variance in mate access will be determined by dominance. Female territoriality will reduce the number of females that occupy an area, and consequently restrict access by males.

Female benefits from these systems can also be arrayed along a number of axes (Figure 4.2). The prospect of direct male assistance is weakly negatively correlated with the axis for male mate access, as males will be less capable of assisting the progeny of several females than those of a single female. Protection from male interference such as infanticide will be higher where males are territorial, although Getz and Hofmann (1986) report high numbers of visits to nests by non-territorial males in the monogamous *M. ochrogaster*. The final axis of variation is the opportunity for mate choice by females. Choice is probably least possible where males are territorial, particularly where the male

Figure 4.1: Hypothethical effect of mating system on the assurance of paternity and access to mates by males. Female territoriality will reduce male access to mates and assurance of paternity

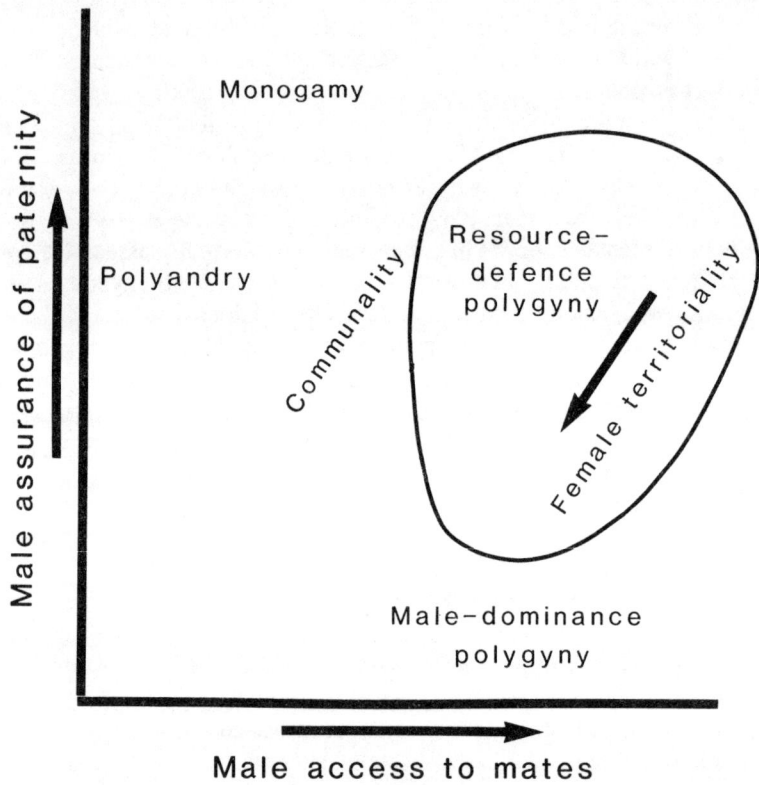

territory is much larger than the home range of the female. The benefits gained by males and females in communal systems has not been ascertained, and classification is uncertain.

Operational Sex Ratio and the Mating System

One obvious clue to mating systems is sex ratio among breeding adults (operational sex ratio, or OSR). This measure provides a simple measure of the potential for polygyny (Clutton-Brock, Harvey and Rudder 1977; Emlen and Oring 1977), though whether bias in OSR is a cause or consequence of different mating systems remains a topic of debate (Murray 1984). Within microtine species, some extreme disparities from an equal number of males and females have been recorded. For example, there were usually about four resident females to every male in a population of *Microtus oregoni* studied by Redfield *et al.* (1978b). As the sex ratio of newly captured animals was much closer to parity

Figure 4.2: Hypothethical effect of mating system on the extent of interference and access to assistance by males to females. These factors will favour the evolution of monogamy where interference is so intense that male investment will consistently be frustrated

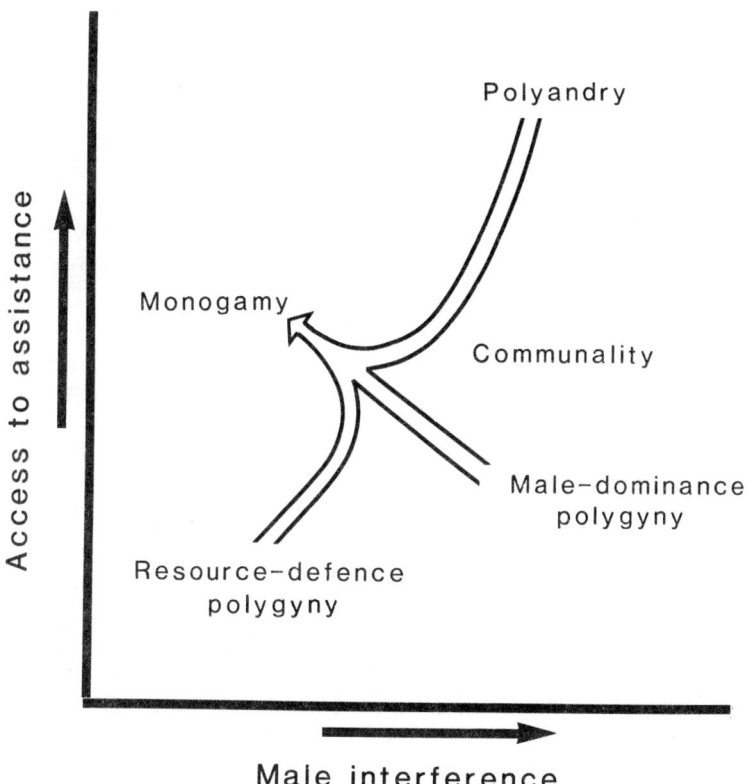

(about 44 per cent male), it seems that spacing behaviour by resident males may force young males to disperse, presumably to their doom. In other microtines, a pronounced female bias in OSR is maintained by distortion of the primary or secondary sex ratio (see Chapter 6). The OSR is density-dependent in some microtines, with a large preponderance of females occurring only at high densities in species where males are territorial and the mating system is resource-defence polygyny (e.g. *M. agrestis* - Hansson 1979c; *M. californicus* - Cockburn and Lidicker 1983; Ostfeld and Klosterman 1986; *M. montanus* - Jannett 1981), but there is no equivalent effect in species where male are monogamous, or show no territoriality (e.g. *M. pennsylvanicus* and *M. ochrogaster* - Myers and Krebs 1971a; *Clethrionomys glareolus* - Hansson 1979c). The correlation between resource-defence polygyny and density-dependence in OSR is both intuitively reasonable, and a promising lead in predicting social systems for species for which demographic data, but not behavioural data, are available.

Several authors have attempted to investigate the effect of sex ratio distortions by conducting removal or supplementation experiments which grossly perturb the sex ratio. It is unfortunate that these experiments are currently confined to species whose social system is poorly known (e.g. *Microtus oregoni*, *M. townsendii*), and those which exhibit female territoriality. In addition, there is little consistency in experimental technique, with methods ranging from total removal of members of one sex (Boonstra and Rodd 1983), repeated removals of a proportion of the members of one sex (Redfield *et al.* 1978a), removals and additions (Redfield *et al.* 1978a, 1978b), or removal of members of one sex at a single period of purported demographic significance (Gipps *et al.* 1985). This confounds interpretation of presence or lack of generality in the results obtained thus far.

The most satisfying result supporting the role of sex-specific social behaviour in population regulation is that, in all but one of the species for which data are available, there is a tendency for females to strongly influence recruitment, survival and sexual maturation of other females (e.g. *Clethrionomys gapperi*- Bondrup-Nielsen 1986; *C. glareolus* - Gipps *et al.* 1985; Bondrup-Nielsen and Ims 1986; *C. rutilus* - Gilbert *et al.* 1986; *Microtus pennsylvanicus* - Boonstra and Rodd 1983; *M. townsendii* - Redfield *et al.* 1978a). The exception is *M. oregoni*, where the control population was highly female biased in any case (Redfield *et al.* 1978b). The effect of female removal on male survival and recruitment is less consistent. Boonstra and Rodd (1983) observed that male *M. pennsylvanicus* survived poorly in the absence of females, a result consistent with dispersal to find mates. Gipps *et al.* (1985) enhanced male survival in *Clethrionomys glareolus* during a spring decline by partial female removal. These results are consistent with the absence of male territoriality in *M. pennsylvanicus* and *Clethrionomys* spp. Redfield *et al.* (1978a, 1978b) found that recruitment of males was lower on both male and female biased grids than on unperturbed control plots in *M. townsendii*, but much higher in *M. oregoni*.

Survival was unaffected by sex ratio manipulation in *M. townsendii*, but survival of the sex which had been removed was much improved in *M. oregoni*. In general, male removal does not influence female survival and recruitment (*Clethrionomys* spp. - Watts 1970; Gipps and Jewell 1979; Bondrup-Nielsen 1986; *Microtus* spp. - Redfield *et al.* 1978a; Boonstra and Rodd 1983), except in *M. oregoni* (Redfield *et al.* 1978b). Male removal does not always enhance male survival and recruitment, and sexual maturation (Watts 1970; Gipps and Jewell 1979; Bondrup-Nielsen 1986), but sometimes dramatic enhancement occurs (*Microtus pennsylvanicus* - Boonstra and Rodd 1983).

Although some of these interspecific differences have been attributed to differences in method (Boonstra and Rodd 1983; Gipps *et al.* 1985), I agree with Redfield *et al.* (1977a) that they are more likely to be indicative of interspecific differences in sex-specific spacing behaviour. Unfortunately the results are not very helpful in illuminating the social systems of *M. oregoni* or *M. townsendii*, apart from suggesting that they are different from each other and from the species which exhibit female territoriality. Extrapolation of this approach to other species will obviously be of considerable interest.

Dominance and Mate Choice

At least in the laboratory, some female microtines prefer to mate with dominant males, while others apparently do not (Figure 3.3). This preference is manifested through either olfactory recognition and preferential association (e.g. Huck and Banks 1982a, 1982b), or females soliciting the presence of the dominant male by calling when approached by a subordinate or unfamiliar male (Webster and Brooks 1981). Familiarity is also partly responsible for mate choice. Shapiro *et al.* (1986) showed that *M. ochrogaster* females mate preferentially and spend more time with familiar males than with unfamiliar males. Familiarity is established by prior copulation, and not just through olfactory contact (Shapiro *et al.* 1986).

Data on the effect of dominance on mate choice are patchily distributed. Dominance influences choice in lemmings, in which the mating system is obscure (Huck and Banks 1982a, 1982b); monogamous species (Shapiro and Dewsbury 1986; Shapiro *et al.* 1986) and species exhibiting mate-defence polygyny (Madison 1980b; Webster and Brooks 1981; Hoffmeyer 1982; Viitala and Hoffmeyer 1985). The correlation of choice with monogamy and mate-defence polygyny is a pleasing vindication of expectations, as females will have the greatest opportunity to choose where male ranges are small relative to their own, or where male ranges overlap extensively. The absence of choice for dominants in *M. montanus* (Shapiro and Dewsbury 1986; Shapiro *et al.* 1986), where resource-defence polygyny prevails, is also consistent with expectation, as male ranges are much larger than those of females. No data are available for species which nest communally, and those which exhibit resource-defence polygyny and lack female territoriality. The latter species should resemble *M. montanus*, but it is not possible to predict the behaviour of communal systems until more is known of the variation in male reproductive success in these systems.

Copulatory Behaviour

Copulation in voles involves three types of behaviour:
(1) mounting without insertion;
(2) intromission with thrusting and insertion but no ejaculation; and
(3) intromissions with a variable number of ejaculations.

I have already mentioned that males may use prolongation of intromission as a means of guarding their mates from other males. It might be anticipated that this behaviour should be most pronounced in species where interference by other males is most likely. However, while prolongation of intromission through several ejaculations appears to be of variable importance, the correlation with mating system is obscure. There appear to be phylogenetically based differences in the copulatory behaviour of *Clethrionomys* and *Microtus* (Milligan 1979), and data for other genera are too fragmentary to permit comparison.

There is considerable interspecific variation in *Microtus* and some evidence of correlation with mating system. Species where males are territorial, but females are not, exhibit few intromissions before ejaculation, many thrusts per intromission, but low ejaculatory frequency (Figure 4.3; but see anomalous data for *M. agrestis* in Milligan 1975). The low ejaculatory frequency would be consistent with low probability of kleptogamy, as female ranges will be normally encompassed by those of the male. Repeated thrusting probably serves to induce ovulation through vaginal stimulation (Kenney, Lanier and Dewsbury 1977). Monogamous species are intermediate with respect to intromission frequency, have a low number of thrusts per intromission, and few ejaculations (Figure 4.3). Once again this is consistent with a low probability of kleptogamy. In addition, males ignore alternative mates (Fuentes and Dewsbury 1984) and there is no Coolidge effect (stimulation by novel females of copulation in satiated males; Gray and Dewsbury 1973).

The remaining species for which data are available are *Microtus pennsylvanicus* and two close relatives *M. montanus* and *M. canicaudus*. All exhibit female territoriality, but there is no male territoriality in *M. pennsylvanicus*. These species have a large number of intromissions prior to ejaculation, a high ejaculation frequency, and a low number of thrusts per intromission (Figure 4.3). These traits are consistent with the high probability of kleptogamy in the male dominance polygyny exhibited by *M. pennsylvanicus*. However, the result for *M. montanus* is surprising, as both males and females are territorial, and assurance of paternity should be as great as in the monogamous and resource-defending species. A possible explanation for this inconsistency is the very low levels of association between males and females in *M. montanus*. Jannett (1982) used radiotelemetry to examine nesting associations in this species and observed only transitory associations between breeding adults. In addition, *M. montanus* scored lowest in Dewsbury's (1985a) classification of the amount of time parents spent at the nest with their pups in the absence of a mate (Figure 4.4). Even though facultative monogamy may prevail in *M. montanus* populations at low densities (Jannett 1981, 1984), it seems unlikely that paternal care and involvement ever occurs in this species. Even male *M. pennsylvanicus*, with a very low assurance of paternity, show some propensity to care for their pups in the laboratory (but see Oliveras and Novak 1986). It is certainly true that the difference between *M. montanus* and the species with male but not female territoriality justifies my recognition of two resource-defence polygynous systems.

Although this discussion has primarily been couched in terms of male advantage, the data of Kenney, Hartung and Dewsbury (1979) support the idea that several ejaculations are necessary to ensure fertilisation, even in the absence of sperm competition. Therefore, female benefit may have a more critical role in

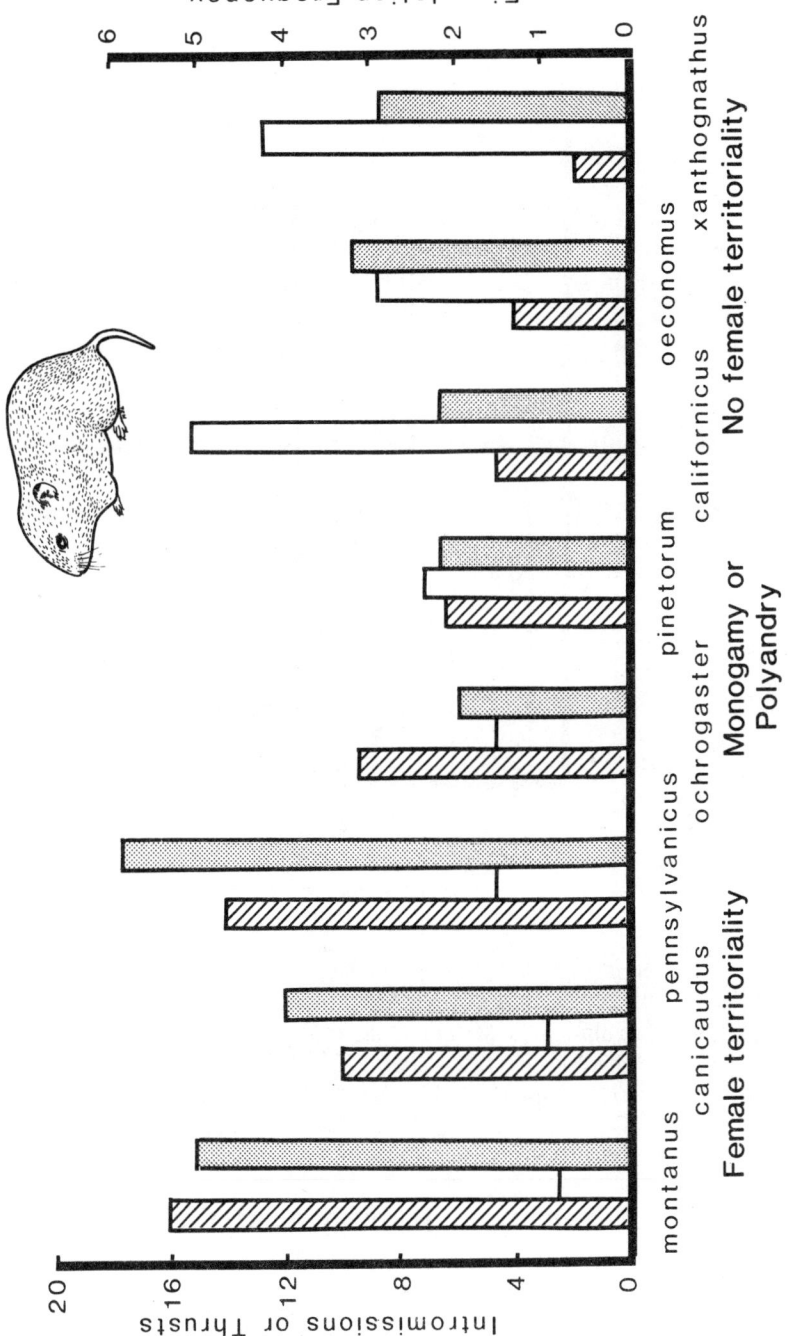

Figure 4.3: Number of intromissions per mount (slanted lines), number of thrusts per intromission (open bars), and number of ejaculations per copulatory sequence (stippled lines) in eight species of Microtus. Data from Kenny, Hartung and Dewsbury 1977; Dewsbury and Hartung 1982

Figure 4.4: The amount of time female and male parents spend on the nest in one 30 minute period when the other parent is absent in four species of Microtus. *After Dewsbury (1985a)*

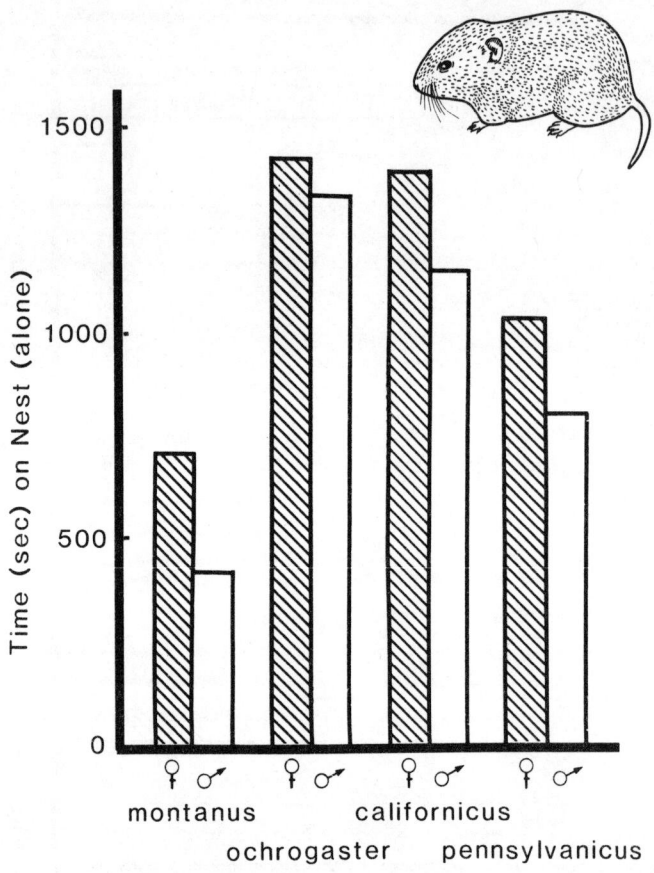

the evolution of these behaviours than has been generally acknowledged (Doty 1974). In particular, the necessity for prolonged vaginal stimulation in the induction of ovulation in *M. montanus* may be implicated in the copulatory behaviour of this species (Davis *et al.* 1974; Gray *et al.* 1974, 1976; Kenney and Dewsbury 1977). Infanticide is unlikely to be implicated, as presence of the stud male can maintain pregnancy in this species (Berger and Negus 1982).

Causes of Variation: Habitat or Demographic Models?

Theoretical predictions of the habitat characteristics promoting the evolution of different mating systems have met with varied success (Emlen and Oring 1977; Wittenberger 1979a, 1981; Oring 1982; Vehrencamp and Bradbury 1984). One outstanding problem is that explanations are often derived *a posteriori* and predictions of correlations between habitat and strategy are rarely unique to a single adaptationist hypothesis. For example, it has been suggested that the outcome of competition for mates in an explosive breeding assemblage may lead to mate-guarding monogamy (in ducks), and to sequential polyandry (in phalaropes) (Wittenberger 1979a; Oring 1982).

For microtines, the only comprehensive theory of mating system evolution is the account of territoriality developed by Ostfeld (1985a). He arrays the different systems as a continuum between female territoriality and male overlap on one hand, and male territoriality and female overlap on the other. Systems where both sexes are territorial are regarded as intermediate, with polygyny developing where female ranges are extremely small. As already explained, an axis of renewability of food resources is the principal determinant of this classification. He pays no attention to the communal and highly variable systems of the type reported in *M. arvalis and Lemniscus*.

I do not believe that the diversity of mating systems can be arrayed along a single resource axis. In the following section I deal with each of the mating systems, and then attempt a general model for the evolution of microtine mating systems.

Monogamy

The unusual case of *Microtus ochrogaster* has attracted most attention within the recent literature on social evolution in microtines. Monogamy among mammals is unusual, but appears to have arisen in response to a variety of selective pressures (Wittenberger 1979a; Wittenberger and Tilson 1980; Dunbar 1984). Kleiman (1977, 1981) suggests that many cases of monogamy in mammals are a reflection of ecological constraints. In particular, if females are sufficiently widely dispersed to frustrate male attempts to defend more than one female territory, *facultative monogamy* will ensue. Under these circumstances polygyny should occasionally occur, and females need derive little benefit from male assistance. By contrast, if the probability that a female can raise offspring without male assistance is low, then strong pair bonds may arise, and paternal care should be better developed (*obligate monogamy*). There should be a continuum of behaviour between these extremes (Kleiman 1981; see Table 4.3).

Because female *M. ochrogaster* seems not to possess any characteristics which make them more dependent on paternal care than other microtines, the main hypothesis proposed to account for this extraordinary occurrence implicates dispersion of females, and as a corollary, facultative monogamy. As the prairie habitat occupied by *M. ochrogaster* is hypothesised to provide *comparatively* constant conditions and promote even dispersion, and because this species requires forbs which have to be harvested over a large area, stable monogamous relationships between males and females have been predicted (Getz and Carter

1980; Getz and Hofmann 1986). However, although there is general support for the suggestion that home range size in mammals should reflect the difficulty of finding and harvesting food (Mace and Harvey 1983), there is no evidence that there is a significant difference between home range size in *ochrogaster* and other microtines (Table 4.4). While any conclusions are confounded by the unusual habitat in which many species are studied, the data do not support the idea that female territory size is unusually large in monogamous species. The additional inconsistency of this hypothesis of constant conditions with the dramatic irruptions exhibited by *M. ochrogaster* has led Getz (1978) to suggest that a breakdown of conditions which facilitate pairbonding may precipitate outbreaks of this species. The circularity of this argument does not require elaboration. Further, although the frequency of monogamous pairs declines slightly with increasing density, monogamy remains the modal mating system, contrary to Getz's speculation (Table 4.3).

Although the diversity of social conditions in which *M. ochrogaster* is found weakens the evidence for obligate monogamy, its importance seems amenable to experimental tests using procedures well developed in analyses of population regulation in other microtines. Redfield *et al.* (1978a, 1978b) demonstrated that it was possible to change the sex ratios of *M. townsendii* and *M. oregoni* towards a male or female bias using repeated removal of one sex. The consequences of removal experiments in different species are dissimilar, raising interesting questions about the differences between the species in the extent and the way in which females regulate recruitment. In the case of *M. ochrogaster*, the predictions are simple. Where a population is manipulated towards a female bias, the fertility rates of females should not decline significantly if monogamy is facultative. In the case of obligate monogamy, a proportion of females should remain unmated, or the capacity of the females to produce young should be impaired. Appropriate experimental protocol has been

Table 4.3: Percentages of breeding units of *Microtus ochrogaster* consisting of a male-female pair (monogamy), a single female only, and more complex breeding units (from Getz and Hofmann 1986). The nature of the complex associations is not made clear in the original account, but they included cases where both more than one male and more than one female were present

	No.	Monogamy	Single ♀	Complex
Overall	281	50.2	27.0	22.8
Low density	72	70.8	16.7	12.5
High density	209	43.1	30.6	26.3
Winter	61	41.0	4.9	54.1
Other	220	52.7	33.2	14.1
Breeding	203	53.2	34.0	12.8
Non-breeding	40	52.5	17.5	30.0

Table 4.4: Home range sizes for microtine species. It should be noted that home range may vary with density in some species, and is dependent on the configuration of habitat. For example, the ranges of *Arvicola terrestris*, *Microtus pinetorum* and *M. richardsoni* are essentially linear. Estimates are dependent on the method of calculation, with trapping estimates tending to underestimate range and statistical treatment of home ranges derived by telemetry tending to overestimate home ranges (Samuels *et al.* 1985). I have preferred data based on telemetry, and data from the spring or summer (see Madison 1985 for sources of variability) (Te = telemetry; Tr = capture-mark-release studies; Ra = radioisotopes)

	Range size (m^2)		Method	Habitat	Source
	Male	Female			
Polyandry					
M. pinetorum	45	42	Te	Orchard	FitzGerald and Madison (1983)
Monogamy					
M. ochrogaster	227	144	Te	Field	Gaines and Johnson (1982)
Resource-defence polygyny (females territorial)					
M. montanus	—	11	Ra	Field	Jannett (1978)
Resource-defence polygyny (females non-territorial)					
M. agrestis	600	300	Tr	Field	Myllymäki (1977b)
M. californicus	129	85	Te	Field	Ostfeld (1986)
M. oeconomus	2,500	530	Tr	Natural	Tast (1966)
M. xanthognathus	650	583	Tr	Natural	Wolff and Lidicker (1980)
Male dominance polygyny (females territorial)					
C. glareolus	4,120	1,280	Tr	Spruce	Korn (1986)
C. rufocanus	1,170	271	Tr	Hedgerow	Kawata (1985a)
C. rutilus	1,313		Tr	Natural	Viitala (1984)
M. pennsylvanicus	192	68	Tr	Field	Madison (1980b)
M. richardsoni	770	222	Te	Natural	Ludwig (1984)
Communal systems					
M. arvalis	900	375	Tr		Nikitina *et al.* (1972)
Mating system unclear (unlikely to be monogamous)					
M. oregoni	907	575	Tr		Gashwiler (1972)
M. townsendii	900	500	Tr	Field	Taitt and Krebs (1981)

developed by Hannon (1983, 1984) in her investigation of social evolution in ptarmigan (see below).

One interesting demographic anomaly in *M. ochrogaster* is an unusually high frequency of winter and year-round breeding, at least in part of its geographic range (Keller 1985), and I hypothesise that this may be involved in the evolution of monogamy. Rose and Gaines (1978) suggested that reproduction in populations in Kansas was intermediate between that of dry- and cold-adapted microtines, and although reproduction declined in midwinter and midsummer, animals bred year-round to overcome these periods of seasonal adversity. In *M. pinetorum*, which has the social system most closely resembling that of *M. ochrogaster*, year-round breeding has also been reported (Paul 1970; Cengel, Estep and Kirkpatrick 1978; but see Valentine and Kirkpatrick 1970), and Lepri and Noden (1984) report lack of sensitivity to photoperiod in males. This is in sharp contrast to many other microtines, including representatives of almost all other mating systems (e.g. *M. agrestis* -Grocock 1979, 1981, Clarke 1981; *M. arvalis*- Dobrowolska and Gromadzka-Ostrowoska 1983; *C. glareolus* - Tähkä *et al.* 1983, Kruczek 1986; *M. montanus* - Horton 1984; but see *M. californicus* - Nelson *et al.* 1983).

One consequence of year-round breeding may be a much more complex cohort structure than prevails in populations with a more limited breeding season. Synchrony in breeding is less likely, and therefore it seems possible that in contrast to the model proposed in Figure 4.2, unprotected females may be particularly subject to interference from males and other females which are not preoccupied with reproduction (Yom-Tov 1975). There is no strong evidence for infanticide in field populations of prairie voles (see Heske and Nelson 1984). However, a number of observations support this proposition. First, visits by strange males to nests of unpaired females are very common in the field (Table 4.5). Second, agonistic encounters between females and unfamiliar sexually-experienced males are frequently observed (Getz, Carter and Gavish 1981; Gavish, Carter and Getz 1983), suggesting that fighting could also provide a behavioural barrier to reproduction and infanticide. I have already mentioned that females prefer to mate with a male with whom they have previously copulated

Table 4.5: Average time between visits by non-resident *Microtus ochrogaster* to the nests of breeding units of different complexity (data from Getz and Hofmann 1986). The difference between female and male visits is significant, as in the high frequency of visits by non-resident males to the nests of single females

Type of breeding unit	Time in days between visits by	
	Males	Females
Monogamous	5.5	6.6
Single female	2.4	5.9
Complex	4.6	6.9

than with an unfamiliar male. Last, the monogamous and polyandrous species of *Microtus* show unusually prolonged sensitivity to pregnancy block in the presence of a strange male (Stehn and Richmond 1975; Kenney, Evans and Dewsbury 1977; Schadler 1981, 1985). In *Microtus pinetorum*, not only are males infanticidal in the laboratory, but lactation does not suppress abortion in response to strange males, as is the case in other rodents for which the Bruce effect is known (Schadler 1985). Because the Bruce effect is now commonly interpreted as a prophylactic against infanticide (Chapter 5), this extension of effect provides strong evidence for the selective importance of infanticide in monogamous species.

Male investment of paternal care in *M. ochrogaster* may therefore be interpreted as protection of investment in the face of interference by other males. In addition, because demographic complexity will accentuate the variance in male attributes, females may have the opportunity to choose between alternative males according to the probability that they will provide care. This may accentuate the difficulty that males will have in cheating on this system. Although I believe that my hypothesis is preferable to those based on food dispersion, several problems are immediately apparent. First, there may be no generality in the year-round breeding of *M. ochrogaster* (Krebs, Keller and Tamarin 1969), and interpretation of which pattern is most representative of the conditions in which monogamy evolved, and its current utility, must rest on plausibility arguments of dubious scientific merit. Second, whether breeding asynchrony influenced the evolution of *M. pinetorum* is still unclear. Field workers appear to prefer to observe this species in apple orchards, no doubt stimulated by the availability of funding (FitzGerald, pers. comm.). However, aspects of my hypothesis do seem testable with both descriptive and experimental studies of both species, and I hope critical examination of these hypotheses will be encouraged by this speculation.

Cooperation and Cohesion, and the Development of Communal Systems

In common with much of modern behavioural theory, conflict over resources and mates has received overwhelming emphasis in studies of microtine behaviour. In this section I review recent evidence that cohesive aspects of social organisation have been underestimated, and attempt to interpret the significance of the communal patterns exhibited by *Microtus arvalis* and *Lemniscus curtatus*, and, to a lesser extent by many other species, such as *Microtus pinetorum*, where reproductive males associate with individual females.

Group nesting.
It is now clear that many microtine species nest communally in winter, apparently to reduce energy expenditure (e.g. Grodzinski *et al.* 1977; Madison 1984). It had been thought that these nest groups disintegrated with the commencement of reproductive activity and establishment of territories by the

territorial sex. However, radiotelemetry has revealed communal nesting in the breeding season in species with very different mating systems (e.g. in male territorial *M. californicus*, R. Ostfeld, pers. comm.; in communal dwelling *M. arvalis*, Frank 1954, de Jonge1983; C. Boyce, pers. comm.; in polyandrous or monogamous *M. pinetorum*, FitzGerald and Madison 1983; Jannett 1978, 1980; in monogamous *M. ochrogaster*, Getz and Hofmann 1986) but whether more than one female within a group becomes reproductive has remained unclear. McShea and Madison (1984) have now reported extensive communal nesting among reproductive female territorial *M. pennsylvanicus*, and confirmed that in at least two cases two litters were in the nest. The extent of this communality declined as the spring progressed, leading to speculation that increasing temperatures reduce the probability that females would form new alliances if their nesting partner died. The occurrence of this behaviour in a species exhibiting female territoriality suggests that it may be worth pursuing in many other species.

McShea and Madison (1984) also speculate that the nesting associations will mainly be established between relatives which have nested together during the winter, and mothers may allow any pups within the nests to suck. In a related observation, I have observed that female *Antechinus stuartii* (a small dasyurid marsupial), also form nesting associations during the period they are lactating and have pups in the nest. These associations are of two sorts. The first is rare and occurs between relatives who share a nest with their respective young. I am currently unable to determine whether mothers suckle their relatives' pups. The second, more common type of association is between mothers who nest together in trees which do not contain offspring. Each mother in the association (up to four females) has a private nest which she visits only to suckle her young. These observations are perplexing because the presumed thermoregulatory benefit to the mother will be countered by the thermoregulatory costs to her offspring. These costs should be particularly severe in small marsupials which acquire endothermy only slowly (Lee and Cockburn 1985a). The most likely explanation is that the mother avoids her young to regulate the costs of lactation, and suckles her young in a solitary nest to minimise the risk of predation.

The more extreme manifestations of this behaviour are the persistent aggregations observed in the more social species, where true group territoriality also appears to occur (e.g. *Microtus pinetorum*, FitzGerald and Madison 1983). It is possible to elaborate models which predict the evolution of territoriality in individuals to the case of group territoriality (Brown 1982). The cost of defence of a territory will be a function of its attractiveness to intruders, and also of population density. Both the patchiness of the environment and the state of the cycle are therefore implicated in interpretation of microtine sociality. Brown's model predicts the point at which sociality will evolve but assumes that new additions to a group become subordinate to the original owner. According to this view, the dominants will only admit additional members to a group when they gain by doing so, but there may be conflicts of interest between dominants and subordinates (e.g. Davies and Houston 1983). Brown (1982) claims that group territoriality, in contrast to more relaxed social associations such as winter aggregations, is inevitably associated with *reproductive* aggregations of three or more individuals. He also suggests that it is unknown in highly fecund species whick lack male parental care and/or protection of the young. The microtine examples urgently require further study.

Environmental Potential for Polygyny.

Emlen and Oring (1977) have proposed that the opportunity for polygyny will depend on the spatial and temporal distribution of resources. The dispersion of mates is likely to be restricted in both space and time (Chapter 2). The potential for polygyny is greatest when resources are spatially clumped and where the availability of mates in time is intermediate between high synchrony and stochastic and unpredictable (Figure 4.5).

Ostfeld (1985a) is clearly correct in suggesting that where females are territorial they will be more evenly dispersed, and consequently it will be more difficult for males to exclude other males from the area occupied by the female. All but one of the cases of resource-defence polygyny occur where females are not territorial. The exception is *M. montanus*, where female home ranges appear to be extremely small (Table 4.4), and hence unusually easy to defend.

Male-dominance polygyny is predicted to occur when mates or other critical resources cannot be monopolised. There is no evidence that female ranges in the species exhibiting male-dominance polygyny are smaller than in species exhibiting resource-defence polygyny, or that they overlap less (Table 4.4). The linear home ranges of stream-dwelling *Arvicola terrestris* and *Microtus richardsoni* provide a constraint of possible importance. Defence of the extremes of the range require regular traversing of the entire range area, and might be economically impossible. In short stream tributaries only one end may need to be defended, and Anderson *et al.* (1976) reported finding only a single male in these circumstances.

The position of *M. pennsylvanicus* is least easy to explain within this framework, unless we accept Ostfeld's argument that the dicotyledons on which these voles depend have lower renewal rates than the diets of other *Microtus* spp., and hence are an undefendable resource. Getz (1978, 1985) has suggested that the behaviour of this species is consistent with its evolution in small, ephemeral habitat patches characterised by instability and the necessity for prolific dispersal (see also Batzli *et al.* 1977). His analysis was almost certainly clouded by early overestimates of the frequency of monogamy in *M. californicus* (see Lidicker 1979, 1980), which strengthened an apparent correlation between monogamy and habitat configuration among species of *Microtus*. Even if Getz's (1985) habitat classification is accepted, it fails to distinguish between species exhibiting the two forms of polygyny (see also Böstrom and Hannson 1979; Hansson 1984c).

Figure 4.5: Factors influencing the environmental potential for polygyny (EPP). Polygyny is favoured when the cloud of points is most dense, when resources are clumped and when mate availability tends towards asynchrony. Modified from Emlen and Oring (1977), and Oring (1982)

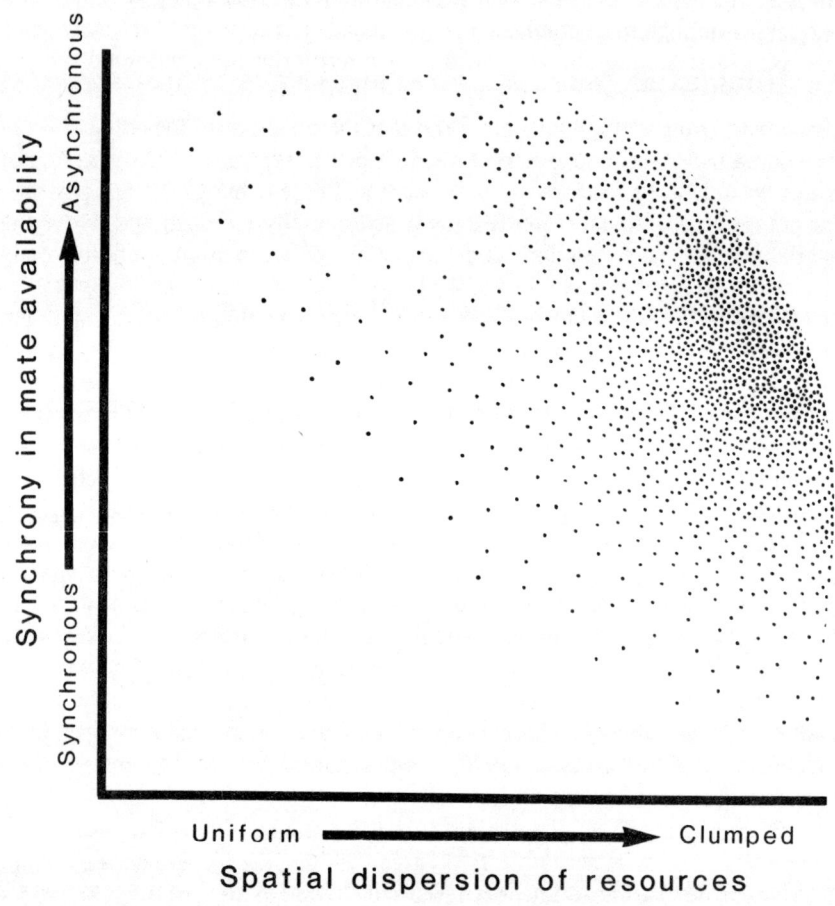

A General Model of Mating Systems in the Microtines

In summary, the mating options available to microtines are constrained by a number of temporal and spatial constraints. Where oestrus is highly synchronised, there may be intense competition for mating opportunities, and promiscuity may ensue, even at the expense of future reproduction. This is most likely to occur with the onset of reproduction in highly seasonal environments, and may partly account for the spring decline observed in some rodents

(Lee and Cockburn 1985b). Where opportunities for reproduction are less tightly constrained by seasonal factors, loss of synchrony may initially increase the potential for polygyny. Where females are highly clumped and female ranges overlap and the boundaries are contiguous, resource-defence polygyny will develop. However, in species where females are territorial, polygyny should be expressed through male dominance unless female territories are extremely small. Defence of space may be very costly when territories are configured along a stream. Where male dominance is strongly developed there should be selection for female choice to minimise the risk of infanticide by the dominant male, and selection for male copulatory strategies which mimimise the risk of sperm displacement. Where the pressure of marauding males is very great, there may be selection promoting mate guarding by males. In extreme cases, it may be necessary for males to supply paternal care to ensure paternity and successful reproductive investment. The tendency towards monogamy may be augmented by increased female susceptibility to the effect of strange males, manifested through abortion or resorption. Such increased susceptibility may arise because the presence of a strange male may indicate to the female the loss or inadequacy of her mate. Circumstances favouring these developments may be the demographic complexity of year-round breeding, which will promote both asynchrony in female reproduction and increased variance in male quality. Monogamous species should therefore have a more pronounced tendency towards female choice than in other systems where males are territorial and occupy very large territories. Such territoriality would normally be expected to reduce the mating options available to females, by increasing the dispersion of males.

The biology of communal systems is not currently understood. Boyce and Boyce (1985) provide the tantalising observation that the development of communality may be both density- and frequency-dependent in *Microtus arvalis*, and consequently this species may be very useful in experimental analysis of the interaction between population and social dynamics. Future publications will be awaited with great interest.

Social Behaviour in Grouse and Hares

An equally comprehensive analysis of territoriality and mating systems in grouse, ptarmigan, and cyclical species of hare is unlikely to elucidate the relationship between population and social dynamics any further, because of the absence of any strong correlation between demography and mating system within the microtines. However, a few brief comments are warranted to reinforce the themes developed throughout this chapter.

Grouse and ptarmigan.

The mating systems of grouse and ptarmigan are also characterised by diversity, including the bizarre lekking behaviour of some species (Wiley 1974; Wittenberger 1978, 1979a; Oring 1982). The cyclical species differ from most voles in that they normally form monogamous pairs which jointly defend a territory against intruders during the breeding season, in contrast to the polygyny and promiscuity prevalent in some of the non-cyclical species (Bergerud and Mossop 1984).

Perhaps influenced by studies of other birds, studies of territorial behaviour have emphasised male behaviour. Substantial evidence has accrued that male territoriality limits access of surplus males to mates (Watson 1967; Watson and Jenkins 1968). Size of male territories appears to reflect both intensity of intruder pressure, and the level of aggressiveness of territorial cocks (Pedersen 1984). Mate-guarding by male willow ptarmigan is pronounced, but may be a response to very high levels of nest predation, and appears designed to ensure paternity in renesting attempts (Martin 1984). Hannon (1983, 1984) has experimentally demonstrated that in some ptarmigan populations, female territoriality also limits the density of breeding hens (see also Lewis 1984). Her data are particularly important, as they also showed that numbers of females and males did not appear to be influenced by reductions in density of the opposite sex. Her research protocol would be helpful in elucidating the mating system of *Microtus ochrogaster*. These conclusions also suggest that both sexes exhibit territorial behaviour, a pattern in distinct contrast to the modal condition among microtines.

Snowshoe hares.
Research on the social behaviour of cyclic hare species is in its infancy. Nonetheless, sufficient is known to suggest that there are no common patterns between microtines, hares, grouse and ptarmigan. First, there is evidence that neither sex is territorial, though animals of both sexes occupy permanent but overlapping home ranges (Boutin 1980). Unlike voles and grouse (e.g. Watson and Jenkins 1968; Krebs *et al.* 1976), removal of females from an area does not prompt other females to shift their home ranges or settle from outside the study plot (Boutin 1980). Removal of adults during the breeding season does enhance recruitment and survival of offspring (Boutin 1984b). Adult removal affects female survival most dramatically in the summer and male survival most in the autumn. Juveniles disperse at an early age (Boutin 1984b), and at all densities (de Poorter 1984).

As might be expected, dominance and aggressive interactions are pronounced in a species with no territoriality. Graf (1985) showed that hierarchies were dominated by males during the winter, but females during the summer. This reversal is consistent with other observations of aggression in breeding females (Grange 1932; de Poorter 1984). De Poorter (1984) also contends that it is unlikely that snowshoe hares can recognise kin and reports an observation of two females suckling the same litter.

Summary and Future Prospects

The use of radiotelemetry and removal experiments to elucidate the mating systems and patterns of recruitment in microtines has revolutionised our interpretation of their social behaviour. Available evidence does not support the hypothesis that common aspects of spacing behaviour underly population fluctuations in microtine rodents. Instead, microtines with similar demography are characterised by diversity in mating systems. This difference appears in part to be a reflection of differences in patterns of territoriality. The mating systems of other vertebrate species which fluctuate in abundance are also sufficiently dissimilar from the various types exhibited within the microtines to weaken support for the notion that gross patterns of demography influence the evolution of sociality.

Topics for Future Research

(1) It is becoming clear as research on mating systems progresses that the cohesive aspects of microtine sociality have been greatly underestimated by students of population regulation.

(2) Ostfeld's model of food renewal currently affords the best explanation of the distribution of female territoriality, and hence of the form of polygyny exhibited within the microtines. Despite this we have no empirical data on food renewal rates. As these will vary with grazing and browsing intensity, with plant species and between habitats, experimental and descriptive data are urgently required.

(3) One of the factors hindering analysis is the highly fragmented data set, frustrating comparisons within and between mating systems. While I have little doubt that some of the deficiencies will have been rectified by the time this book is published, I have attempted to summarise the pattern of research in Table 4.6. The gaps are topics ripe for the plucking, and can often be achieved with comparatively little effort - though unfortunately not by an Australian!

Table 4.6: A summary of the information available on voles according to their mating system. The existence of polyandry in the microtines remains questionable. (However, such a system has been assumed for *Microtus pinetorum* — see text)

Mating system	Polyandry	Monogamy	Resource-defence polygyny		Male-dominance polygyny	Communal systems
			Females territorial	Females non-territorial		
Best studied species	*M. pinetorum*	*M. ochrogaster*	*M. montanus*	*M. agrestis* *M. californicus*	*M. pennsylvanicus* *C. glareolus*	*M. arvalis*
Female behaviour						
Control of recruitment	?	?	Tentatively yes	Tentatively yes	Yes	?
Choice of dominant males	?	Occurs	Absent	?	Occurs	?
Communal breeding	Rare	Rare	No	Rare	At start of season	Common
Promiscuity	?	?	?	?	Occurs	?
Bruce effect	Prolonged	Prolonged	?	Yes	Yes	Yes
Cost of lactation	Low	?	?	?	High	High
Induced ovulation	Yes	Yes	Yes	Yes	Yes	Yes
Retardation of puberty -urine of grouped females	Yes	?	?	?	?	Unclear
-density-dependent	?	Absent	Present	Present	Present	
Acceleration of puberty -urine of a male	Yes	Yes	?	Yes	Yes	Yes
Breeding dispersal -increasing populations	?	Yes	Yes	Yes	Yes	?

Table 4.6 (continued)

	In field and lab	In field and lab	Absent	Occurs in lab	Occurs in lab
Male behaviour					
Paternal care	Low	Low	High	Low	?
Copulatory frequency					
-intromission frequency	Low	Low	Low	High	High
-thrusts per intromission	Low	Low	High	Low	Low
-no. of ejaculations	No	?	Yes	Yes	?
Photoperiodic effects			Yes	Yes	Yes
Both sexes					
Incest avoidance	Yes	Yes	Yes	Yes	No
Winter aggregation	Aggregate anyway	No	?	?	Yes
Density-dependent OSR	?	No	Yes	Yes	No
Natal dispersal					
-sex bias	Male	No	?	Male	Aggregate anyway ?

Chapter 5
Life History Evolution: Growth and Puberty

Successful reproduction and survival are the currency in which natural selection must be interpreted. Variation in these life history traits is consequently of great theoretical importance, and may be expressed in terms of several important questions. When, or at what size, should an animal first reproduce? How many times should an animal reproduce within a season and within its lifetime? When an animal reproduces how many offspring should it have and what size should they be? If the population is polymorphic or variable, what type of offspring should be produced? Should those young be tolerated or dispersed through space? If reproduction is costly, what level of risk is acceptable within a reproductive bout? As many life history attributes are likely to be correlated, what combination of traits will be favoured by selection? As selection may act throughout the reproductive life of an organism, can selection on prereproductive phases influence the expression of traits in reproducing animals? Abundant theoretical evidence suggests that the answers to these questions are unlikely to be similar for animals living in different habitats, in different social conditions, and in different geographical areas. There is also little reason to believe that all animals within a population should conform to the same pattern, as frequency-dependence is as important in interpretation of some aspects of life history evolution as it is in understanding social interactions (Maynard Smith 1982; Hastings 1984).

In the following chapters, I analyse the factors driving life history evolution in the microtines. However, before tackling the question of density- and frequency-dependence and intraspecific and interspecific diversity in microtine life history patterns, it is worth commenting that the microtines possess characteristics which make them very distinct from most other mammals.

Many life history traits represent a mammalian extreme, but show little variation within the microtine group. In general, gestation is short, females mate post-partum, and microtines have high basal and active rates of metabolism, and do not enter torpor as a way of coping with the extreme climates in which they live. Although subject to considerable variation, litters are large, lives are short and growth is extremely rapid. Some young probably conceive at 15 days, which is a mammalian record. Many of these attributes are unpredicted in small mammalian herbivores (Chapter 2), and all bespeak selection for high fecundity.

Alternative Paradigms for the Study of Density-dependence

Population density plays a central role in life history theory. In Chapter 2, I reviewed classifications of habitats according to the selective pressures they generate. Most of these classifications distinguish between habitats in which populations are continually close to carrying capacity from those in which carrying capacity is rarely attained. Density-dependent effects are likely to be of importance in species which irrupt in numbers more or less predictably, particularly if it is accepted that selection can act on an ecological time scale. The study of density-dependence therefore assumes great significance, and it is worth pausing to consider the methodological difficulties involved.

The Laboratory

It is axiomatic that laboratory conditions are unsuited to the study of density-dependence in mammalian populations, as the range of outcomes of social interactions and attempts at reproduction is inevitably reduced. Laboratory colonies of rodents are subjected to so many departures from 'normal' circumstances that 'controls' in laboratory conditions usually represent a complex multifactorial deviation from the evolutionary framework in which it is usually hoped that data can be interpreted. While laboratory studies are often the only way in which the proximate physiological basis of reproduction and behaviour can be studied, and will be referred to throughout this discussion, there is general consensus that field studies are necessary to understand life history evolution and population regulation.

Use of comparisons of periods of high density and periods of low density within a single population to infer density-dependent effects on social behaviour is fraught with difficulty. Such comparisons do not meet statistical requirements for independent samples (Hurlbert 1984), and are necessarily confounded by effects from any extrinsic variables contributing to interannual differences in density. A similar case can be made for habitats which differ in the density of animals they support at any given time. There are theoretical reasons to believe that animals will assort themselves in a constant patchy environment according to an 'ideal free' distribution, where fitnesses of animals in different habitats are approximately equal, even if the densities in each patch differ substantially (Fretwell 1972). Such assortment and dispersal is, of course, one factor which laboratory studies cannot hope to address.

In order to overcome these frustrating difficulties, a number of methods have been developed which either seek to manipulate density directly, or exploit natural circumstances which alter the nature of population fluctuations.

Removal Experiments

Krebs and his colleagues and students have been the chief advocates of the use of removal experiments to examine density-dependence (e.g. Krebs *et al.* 1976, 1978a; Rodd and Boonstra 1984). Experiments of this sort generate unequivocal conclusions about population regulation. Consistent with predictions made some time ago by Watson and Moss (1970)(Table 5.1), they indicate that spacing behaviour does restrict access to sites for reproduction by young voles (e.g. Krebs *et al.* 1976; Tamarin 1977a, 1983; Gaines, Vivas and Baker 1979; Baird and Birney 1982; Taitt 1985a; Taitt and Krebs 1985) and young grouse (Lewis and Zwickel 1980; Zwickel 1980; Hannon 1983). This conclusion is unaffected by the mating and territorial system of the species involved.

Despite the unanimity and repeatability of this result, the evolutionary interpretations which can be construed from removal experiments is uncertain. While experiments which target particular individuals often provide excellent evidence for causation of behaviour (Chapter 3; Cockburn, Scott and Scotts 1985a), blanket removal of individuals from the population, or even a fragment of the population, has the potential to create such profound social disruption that interpretation of results becomes impossible (Redfield *et al.* 1978b). The ability of short-term density perturbations to contribute to an understanding of evolutionary changes is also questionable.

Table 5.1: Conditions which show that social behaviour limits a breeding population (after Watson and Moss 1970)

(a)	A substantial part of the population does not breed, either because they die, or because they attempt to breed but they and/or all their young die; or because they are inhibited from breeding even though their young survive, and may breed in later years.
(b)	Such non-breeders are physiologically capable of breeding if the more dominant of territorial animals are removed.
(c)	The breeding animals are spaced in such a way that they are not completely using up some resource such as food, space, or nest sites.
(d)	The mortality or depressed recruitment due to social behaviour must change at the same rate and in the opposite sense to other causes of mortality or depressed recruitment. Thus, if mortality is reduced by amelioration of weather or improved resources, then mortality due to social interaction must increase by an equivalent amount.

The Island Model

Many authors have noted that cyclic oscillations are dampened on islands in contrast to the mainland (Lidicker 1973; Gliwicz 1980). For example, *Microtus breweri* is confined to a 2.6 km^2 island off the coast of Massachusetts. It has probably been isolated from its closest mainland relative, *M. pennsylvanicus*, for 3,000 years (Tamarin and Kunz 1974). While mainland voles show "typical" multiannual fluctuations, *M. breweri* does not (Tamarin 1977b). There are several behavioural correlates of this demographic stability, including dispersal tendency (Tamarin 1977a; Keith and Tamarin 1981) and aggressive behaviour (Reich and Tamarin 1980; Reich, Wood, Rothstein and Tamarin 1982). There are genetic distinctions between island and nearby mainland populations (Kohn and Tamarin 1978; Keith and Tamarin; Nygren 1980a, 1980b), but cause and effect are difficult to distinguish (Gaines 1985; Chapter 1).

Strong selection and other evolutionary pressures acting on demographic and behavioural traits in island populations are easy to understand. Populations exhibiting demographic instability are unlikely to persist on very small islands. Pokki (1981) examined the distribution of *M. agrestis* through an archipelago of small islands in the Gulf of Finland. Extinction and recolonisation was extremely common on small islands, and cyclic variation was absent where island size was less than 1 hectare.

Further, chance founder events may profoundly influence the characteristics of island populations, particularly where the species are good colonisers, as is generally true for microtines. Where island populations persist in the absence of density variation, it may be possible to examine the evolutionary differences between populations exposed to fluctuating or stable environments.

Some ecologically important traits show evidence of genetic control. Alder *et al.* (1981) compared mating behaviour in bank voles from mainland Great Britain and the island of Skomer. The mainland form was more prone to initiate mating, exhibited more mounts and intromissions before ejaculation, fought sooner after ejaculation, and had shorter intervals between intromissions. Hybrids were intermediate in tests, but high intromission frequency was dominant in reciprocal hybridisation tests. Data of this sort provide compelling evidence for genetic influences on behaviour, and it is tempting to relate them to the social context of island populations.

While the biology of island populations has proved to be of immense importance in comparative ecology, dissecting cause and effect has been extremely difficult. Founder effects confound adaptative interpretations. A greater problem is that the restrictions on dispersal, habitat heterogeneity, competition and predation which are likely to occur on islands mean that almost any differences between mainland and island populations are amenable to multiple interpretations. This is what has hindered consensus about the factors that influence cyclic populations.

One point that should be made clearly is the many advantages islands offer as experimental and descriptive tools. Because complete enumeration should be tractable, and immigrants unusually easy to identify, more sophisticated genealogical and demographic analysis should be possible than is ordinarily the case (e.g. Bujalska 1985a). Some authors have advocated the use of habitat islands as equivalent research tools (e.g. Nabaglo 1981), but I share Krebs' (1986) scepticism about the certainty with which conclusions about emigration and immigration can be drawn.

Latitudinal Gradients in Multiannual Cycling

I have made repeated mention of the significance for population regulation theory of gradients of cyclicity in both Fennoscandia and North America. Several biologists have sought to use differences in populations from northern and southern populations to indicate genetic differences between northern and southern forms, and to attribute these differences to the selective milieu prevailing in stable and fluctuating populations. Some success has been reported (Table 5.2), but some results show low repeatability. Nonetheless, this is an extremely attractive research paradigm of growing empirical and theoretical influence, and I will be referring to comparisons of this sort throughout the remainder of this text.

Costs, Constraints and Tradeoffs

The model within which life history variation is most commonly interpreted had its origins in the writings of Fisher (1930) and Cole (1954), and has been stated most succintly by Williams (1966b). His Demographic Theory of Optimum Reproduction states:

(1) When reproductives and non-reproductives have the same limited resources available for investment, an increase in *reproductive investment* inevitably results in a decrease in somatic investment.

Table 5.2: Differences between laboratory-bred *Clethrionomys glareolus* derived from southern non-cyclic and northern cyclic populations in Sweden (after Gustafsson 1985)

Characteristics of northern animals	Hansson (1984c)	Gustafsson *et al.* (1983) Gustafsson (1985)
Cyclicity	Yes	Yes
Larger	Yes	Yes
Longer fur, greyer	Yes	Yes
Larger litters	No	Yes
Young survive poorly	Yes	Sometimes
Fewer animals reproduce	Yes	Yes
Greater activity	Yes	Yes
Are differences heritable?	No	Yes

(2) When reproduction takes place at the expense of somatic investment, the somatic costs reduce the probability of surviving to breed again and/or reduce future fecundity.

(3) When reproduction results in survival and/or fecundity costs there is a tradeoff between current reproductive output and *residual reproductive value*.

The Currency of Comparison

The notion that tradeoffs influence the evolution of life history tactics is both pervasive and uncontroversial. The appropriate currency for comparison is much less certain. Much of the early mammalian literature concerned with life history tradeoffs assumes that energy expended on reproduction is proportional to the cost suffered by the organism. The literature on reproduction in small rodents is certainly no exception.

The Cost of Lactation

It is widely assumed that the metabolic costs of reproduction in mammals are borne most heavily by females, particularly during the period of late lactation. Several metabolic analyses are available, but the data are fragmentary and ill-suited to comparative analysis (Table 5.3). Nonetheless, some generalities emerge. In contrast to other rodent genera for which data are available (e.g. *Peromyscus*, microtine energetic costs appear high, and the efficiency of conversion into offspring of the surplus energy intake by females is low. *Microtus pinetorum* diverges markedly from the pattern in other *Microtus* and *Clethrionomys*, with low litter size, high conversion efficency, and low lactation costs. It is tempting to correlate the metabolic patterns with low litter size, and the low litter size with the unusual mating system in *M. pinetorum* (polyandry or monogamy). It needs to be mentioned that there is an allometric relation between cost increments associated with reproduction and body mass, and this can be used to predict that large species (*Sigmodon hispidus* and *Peromyscus floridanus*) should suffer lower costs than small species. Within the small *Peromyscus* spp, the monogamous form (*P. polionotus*, see Foltz 1981) has the smallest cost increments, but not the smallest litter size.

All these studies reflect the costs sustained by a typical female of average litter size. In the analyis of evolutionary tradeoffs, it is probably more useful to ask what are the costs associated with increasing litter size, or what are the benefits of a reduction in litter size. Glazier (1985) uses a graphical analysis to argue that there is an upper limit (five young) to the cost increments associated with increased litter size in small mammals, but his statistical procedures seem completely implausible, being based on the performance of only three species (of a sample of 93). The best data from rodents are for *Sigmodon hispidus*, which shows a latitudinal gradient in litter size associated with recent northward expansion of its range. Mattingly and McClure (1982) compared energy costs of a northern population (mean litter size is 7.0) with two southern populations (4.7 and 5.0). While individuals from populations with large litters did not re-

Table 5.3: The energetic costs of reproduction, and the efficiency of conversion of metabolisable energy in various small rodents

Species	Litter size	Increase over non-reproductives (%)		Efficiency of conversion (%)(see text)	Source
		Gestation	Lactation		
Microtus					
arvalis	4.3	32	133	15.4	1
pennsylvanicus	5.3	36	123	—	2
pinetorum	2.2	—	48	26.4	3
Clethrionomys					
gapperi	5.3	68	144	—	2
glareolus	4	24	92	14.6	4
Peromyscus					
maniculatus	4.3	—	147	21.8	5
leucopus	3.9	—	99	20.9	5
polionotus	3.6	—	74	17.6	5
eremicus	2.4	—	113	20.3	5
floridanus	2.3	—	73	19.6	5
Phodopus					
sungorus	3	10	67	18.3	6
Sigmodon					
hispidus	5	—	39	45	7

Sources: 1- Migula (1969); 2- Innes and Millar (1981); 3- Lochmiller et al. (1983); 4- Kaczmarski (1966); 5- Glazier (1985); 6- Schierwater and Klingel (1986); 7- Randolph et al. (1977)

quire more energy during gestation than the individuals from populations with small litters, there were marked energy costs associated with suckling the larger litters (111 per cent increment versus a 66 per cent increment in ingestion). Animals with large litters also lost weight during reproduction. Both these responses are probably facilitated by the increased body size in northern populations. Under severe food restriction, northern animals lost young until the litter size approached that of southern animals (Mattingly and McClure 1985). These responses suggest that the interactions between body size, reproductive effort and litter size are complex. The evidence of a tradeoff between maternal costs and litter size are obscured by latitudinal differences in body size.

In a related theoretical analysis, Stenseth, Framstad, Migula, Trojan and Wojciechowska-Trojan (1980) have used data on the metabolic rates of pregnant and lactating *Microtus arvalis* to derive a series of tentative conclusions about the significance of metabolic limitations in mortality patterns in microtines. The most interesting result is that the costs sustained by large females are less sensitive to increased litter size than is true for small females. However, this conclusion cannot be used to suggest that large females are insulated from the costs of reproduction, as their metabolic costs are higher to begin with. The maximum litter size at weaning need not depend on the weight or age of the female in a simple way.

Energy is a Misleading Currency

These results raise questions about the validity of the assumption relating metabolic expenditure to reproductive costs. There is a growing consensus that such assumptions are often misleading (Clutton-Brock 1984). Instead, analyses of tradeoffs should reflect the interaction of life history traits such as survival, growth and reproduction themselves. For example, increased litter size reduces the size of the pups and their survival (e.g. Richmond and Conaway 1969; Krohne 1981), but larger mothers may be able to raise a larger mass of young (Myers and Master 1983). In the absence of body size variation, there may be negative correlations between the survival of the mother and the survival of her offspring (Morris 1986), reducing the value of simple one-dimensional measures of metabolic costs.

Reznick (1985) classifies attempts to identify correlations among life history attributes into four classes:

(1) **phenotypic correlations.** the correlation of association of some index of reproductive effort with a potential cost, such as growth, survival or fecundity;

(2) **experimental manipulations.** direct manipulation of some aspect of reproduction or some environental variable which affects reproduction;

(3) **genetic correlations.** quantitative genetic estimates of the correlation between an index of reproductive effort and some potential cost. In these studies the heritabilty of traits and their phenotypic and genetic covariance is examined in order to understand how selection on one life history trait is constrained by other traits;

(4) **correlated responses to selection.** correlated changes in some index of reproductive effort and some potential cost in response to selection on some aspect of the life history.

Reznick (1985) agrees with a growing literature which suggests that analysis of tradeoffs using the tools of quantitative genetics are most useful in understanding evolution of life histories (papers in Dingle and Hegmann 1982; Lande 1982; Rose 1983; Stearns 1983; Lynch 1984; Bell 1984a, 1984b). Despite this enthusiasm, empirical progress is limited, as genealogy is difficult to assess in the field (see however, Price and Grant 1984), and results of heritability studies in laboratory environments are ambiguous and confounded (Service and Rose 1985), particularly when they rely on observation of highly inbred lines (compare Giesel *et al.* 1982 with Rose 1984b).

Growth, Body Size and Age at Sexual Maturity

An obvious component of the high fecundity of microtines is their rapid growth and development, a phenomenon particularly remarkable given the poor nutrition afforded by their diet. Conventional mathematical descriptions of mammalian growth are only weakly applicable to microtine data, apparently because weight does not asymptote (Zullinger, Ricklefs, Redford and Mace 1984). As microtines continue to increase in size after they have reproduced, tradeoffs between growth and reproduction may be particularly important, as is known to be the case in organisms other than higher vertebrates. Patterns of growth and development contribute to two life history traits of unarguable importance. First, an animal which grows rapidly will be larger than other animals in the population, provided that the total growth period for all individuals is similar. Alternatively, an animal which grows rapidly may not attain adult size but may be able to reproduce earlier.

The growth rate of some microtines decreases as litter size increases (Halfpenny and Ingraham 1983). Attempts to correlate growth rates with mating system, habitat and latitude are confused, and confounded by the lack of conformity between microtine growth and mathematical descriptors. *Microtus townsendii* may be used to illustrate the extent of intrapopulation variability. Beacham (1980b) observed that (1) male voles always grow faster than females; (2) spring born animals grow more quickly than animals born later in the reproductive season; (3) the spring born animals are unaffected by density, but animals born later grow more slowly as density increases, though the extent of the effect is influenced by body weight itself. There appear to be differences in growth rate according to levels of 'aggression' (Beacham 1980c) and tendency to disperse (Beacham 1979b), though these differences are also affected by body weight.

Interpopulation effects along geographic gradients are also obscure. Daketse and Martinet (1977) suggested that *Microtus arvalis* grow more rapidly at low temperatures than at high temperatures. Therefore metabolic limitations may be less important than the problems of rapid growth in cold climates.

Interspecific differences are clouded by the lack of an effective growth descriptor (Morrison, Dietrich and Preston 1977; Innes and Millar 1979. Halfpenny and Ingraham (1983) calculated the geometric growth rate according to a formula developed by Brody (1945) where:

$$k = (\ln M_2 - \ln M_1)/(t_2 - t_1) \qquad \ldots 5.1$$

where k is the growth constant, M is mass, and t is time. They present data for nine species which suggest that there is little interspecific variation but *Phenacomys longicaudus* and *Microtus pinetorum* grow somewhat slower than the others. It remains uncertain whether this can be attributed to the poor diet of the former species, or the mating system of the latter. Further comparative analyses are warranted.

Body Size and the Chitty Effect

Variation in body size is easily measured and is assumed to have considerable biological significance. Several authors have commented upon the increase in body size of microtines during the peak of a cycle (e.g. in *Clethrionomys*,

Koshkina and Korotkov 1975; *Lemmus* and *Dicrostonyx*, Krebs 1964a; *Microtus* Boonstra and Krebs 1979; and *Synaptomys* Gaines, Rose and McClenaghan 1977). The change appears to be accompanied by changes in morphology and skeletal proportions, rather than just resulting from an increase in body mass (Krebs 1964b; Andersen and Wiig 1982). However, the ontogenetic nature of these morphological changes remains unexplored, and may have a more simple basis than currently appears to be the case. Boonstra and Krebs (1979) have called this phenomenon the Chitty effect, in honour of the first biologist to document the trend (Chitty 1952). Boonstra and Krebs (1979) attributed this phenomenon to selection, and constructed two scenarios for its origin. The first hypothesis suggested that these animals are r-selected genotypes at an adaptive advantage in increasing populations. The second attributed large size to α-selection in high density populations where aggression is important. The relation between these hypotheses and the Chitty Polymorphic Behaviour Hypothesis should be obvious. Boonstra and Krebs investigated these hypotheses in four species by regressing survival data against population growth rate and density. Surprisingly, they obtained results which favoured the r-selection hypothesis in *M. pennsylvanicus* and *M. townsendii*, and the α-selection hypothesis in *M. californicus* and *M. ochrogaster*. An alternative view has been advocated most forcefully by Ferns (1979), who suggests that nutritional differences underly interannual differences in body weight, regardless of density. His correlative evidence is impressive (Figure 5.1).

At least three other hypotheses suggest that part of the effect is epiphenomenal and does not require adaptive explanation. The first attributes part of the change to changes in age structure (Myllymäki 1977a; Mallory, Elliott and Brooks 1981). The latter hypotheses were influenced by Cockburn and Lidicker's (1983) demonstration that the Chitty effect was most pronounced among animals living in poor habitat unlikely to sustain reproduction by voles (Table 5.4). Ostfeld and Lidicker, who have repeated this result (pers. comm.), have suggested that where reproduction in territorial males is frustrated by absence of access to sexually receptive females, as will be common at high density in species with male territoriality, the costs of reproduction will be absent and this will be reflected in increased or prolonged growth. *Microtus* species are unusual among mammals in that growth is sustained throughout their lives (Zullinger *et al.* 1985). I had earlier proposed that in animals under selection for rapid maturity, reproduction at small body sizes is advantageous and that it may be useful to commence reproduction before growth is complete. This is probably a frequent tactic among slow growing marsupials (Lee and Cockburn 1985a). In poor habitat the nutritional constraints associated with small body size are likely to be felt most, so there may be delayed maturity or selection against small body sizes. Large body size may reduce processing constraints (Cockburn, Braithwaite and Lee 1981; see also Chapter 2). In either case, there will be both a real and artefactual increase in the average body size of the population, particularly at high density. Anderson (1975) could not detect

Figure 5.1: Relation between the standing crop of herbage in a control plot at the Rothamsted Park Grass Experiment, and the average weight of adult Microtus agrestis *from population studies at Lake Vyrnwy (Chitty 1952; Chitty and Chitty 1962), Wytham (Evans 1973) and Darnaford (Ferns 1979). The control plot provides an index of extrinsic factors influencing the availability of food to herbivores. The original compilation was provided by Ferns (1979). I have differed from Ferns in expressing the relation as the reduced major axis (Vole mass in grams = 17.9 + 5.6 (Dry weight of herbage in kilogram × 10^3/ha))*

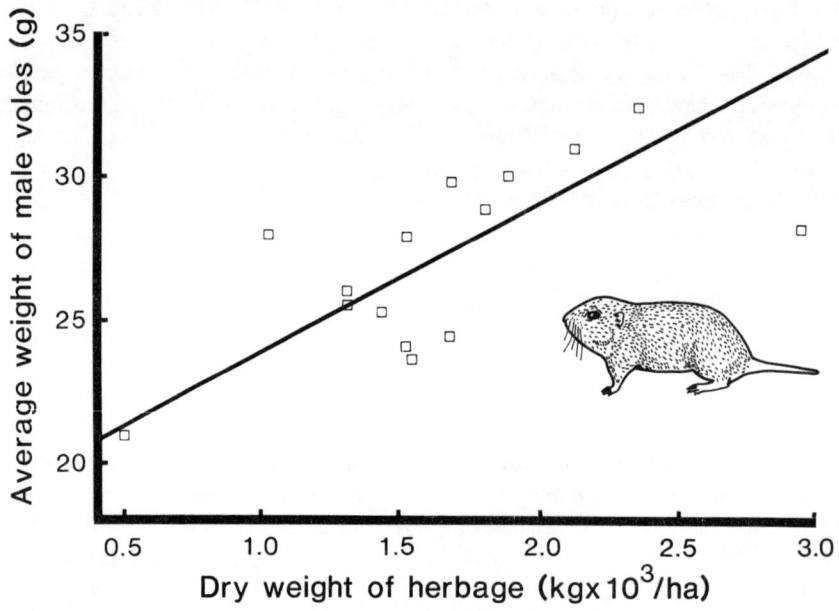

any evidence of heritability for body size in *Microtus townsendii*, and Gustafsson and Batzli (1985) concluded that latitudinal differences in body size in *Clethrionomys glareolus* in Sweden were determined as much by diet, as by any genetic differences.

Whether large body size provides any additional advantages remains extremely confused. If stress-mediated gluconeogenesis has any significance in enabling males to participate in competition for mates (see Chapter 1), then large males may withstand the deterioration in body condition better than small mammals (Scott and Tan 1985). When competition for mates is mediated through dominance (see Chapter 3), then large males may have a better chance of displacing other animals. However, data which support this hypothesis are lacking. Kawata (1985a) observed that there was no correlation between body size and mating success in *Clethrionomys rufocanus*, but there was a weak association between flank gland size and mating success. This suggests that endocrine influences are more important than size alone. In a related observa-

Table 5.4: Spatial variation in the extent of the Chitty effect (weight increase in peak populations) in *Microtus californicus*. Vegetation dominated by the native perennial grass *Elymus triticoides* is preferred by female voles. Reproduction and survival is greatly enhanced in patches dominated by this species. (* denotes the difference in weight at peak population densities was significantly greater than during other population phases — Student-Newman-Keuls multiple comparison test.) Data from Cockburn and Lidicker (1983)

Grid	Vegetation characteristics (classification follows Chapter 2)	% increase in weight Males	Females
1	Dense cover of *E. triticoides* (constant habitat)	−0.6	−3.4
2	Patches of dense *E. triticoides* (constant and colonisation: production)	+9.4	+0.7
3	Predominantly *Conium maculatum* (colonisation: reception)	+13.8*	−1.6
4	Annual grasses (colonisation: production)	+13.5*	+3.3

tion, Hannon and Roland (1984) observed that territory acquisition in willow ptarmigan was not influenced by body size and condition, but was affected by levels of aggression, presumably mediated by endocrine status. The significance of flank glands in microtines requires further investigation. Wolff (1985b) offers a preliminary review. One point relevant to the current discussion is Tamarin's (1981) observation that although glands are normally absent in the populations of *Microtus pennsylvanicus* he has studied, they developed in one old large male, suggesting that endocrine and morphological determinants of dominance may be coupled (see also Jannett 1975; Boonstra and Youson 1982).

I have already mentioned that the metabolic costs suffered by females may vary with their body weight. In some species there is also a clear relation between body weight and litter size (e.g. Tast 1980; Figure 5.2). Because variation in female body weights is extreme (as much as fourfold in Tast's study), and the demographic relevance of increased litter size is clear, it may be unwise to interpret selection for increased size in terms of male advantage alone. Indeed, Gustafsson, Andersson and Westlin (1983) have shown that *Clethrionomys glareolus* from northern cyclic populations have faster growth, larger body size and larger litters than animals from southern non-cyclic populations. The evolutionary implications of these differences in litter size will be addressed below.

Social Retardation and Acceleration of Reproduction

Unlike changes in body size, the importance of changes in the period over which

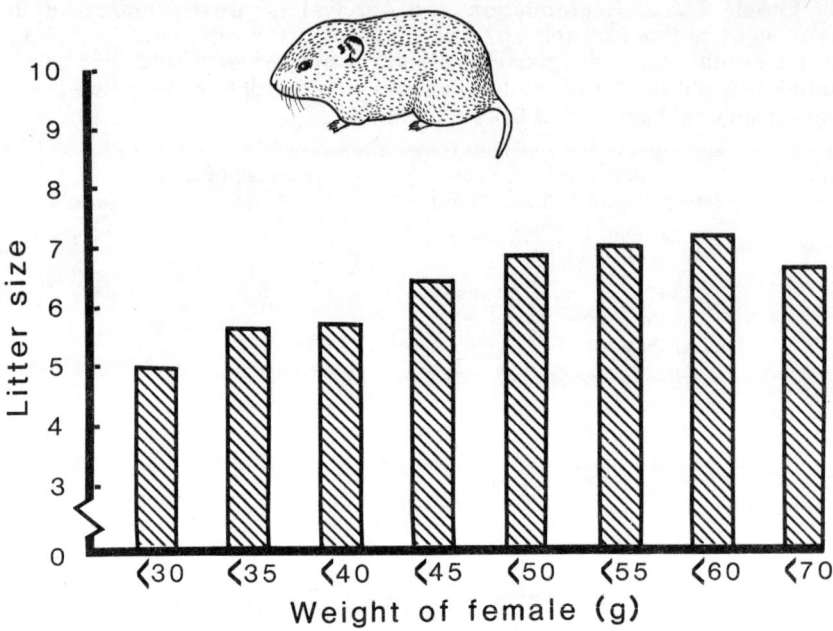

Figure 5.2: *Litter size versus weight of female* Microtus agrestis. *Data from Tast (1980)*

development takes place seems quite clear. The age at which reproduction commences is the most sensitive of parameters in determining the rate of population growth and the efficacy of life history strategy, particularly in expanding populations (e.g. Mertz 1971; Stearns 1976). This is because the multiplicative effect from the reproductive contribution of an animal's descendants is felt earlier when individuals mature quickly. Any alteration to onset of reproduction is therefore of particular interest.

Some species of small rodents show socially mediated acceleration or deceleration of puberty, at least in the laboratory. For example, first oestrus in female house mice is accelerated when they are exposed to unfamiliar males, their urine, or even their soiled bedding. This effect is mediated by peptides in the urine which stimulate gonadotrophin release (Vandenbergh 1969; Drickamer 1983a; Vandenbergh, Finlayson, Dobrogosz, Dills and Kost 1976; Keverne 1983).

The significance of these results for population and social parameters in the field remains unclear. Like many laboratory studies, the social isolation used as a control will normally be a more extreme departure from natural circumstances than exposure to the urine of strange males. Although in controlled experiments the quantity of urine required to produce a significant effect is extremely small

(Drickamer 1984c), the sensory capacity of mice to discriminate cues among the olfactory 'noise' that must prevail in ordinary circumstances remains uninvestigated. However, a number of recent results suggest that there may be considerable adaptive significance for puberty deceleration and acceleration. If the effect is discriminatory in a predictable way, it is much less likely to be an artefact. In the following discussion I first document the nature of the effects which have been recorded in laboratory studies of a variety of rodents. I then discuss the difficulties in reconciling current adaptive interpretation of these patterns in the light of the patterns of social organisation reported in earlier chapters.

Much of the following summary is based on studies of the laboratory mouse (*Mus musculus*), particularly the exhaustive analyses of Drickamer, Vandenbergh and Bronson. Their results are relevant to my discussion because many of their results are repeatable in other species, and house mice also undergo population fluctuations which may be as dramatic as those of the lemmings. However, the ecological position and social organisation of the house mouse may differ substantially from that of microtines, and consequently interspecific generalisations about adaptive significance need to be treated with considerable caution. In particular, house mice are often fugitive species within rodent communities, in contrast to microtines which may displace other species during peaks of abundance (e.g. Berry 1977; Blaustein 1980, 1981; Heske, Ostfeld and Lidicker 1984).

Female-female interactions.

Females exposed to the urine of other females which have lived in groups for more than ten days show retardation of puberty in contrast to females kept in isolation (Drickamer 1979; 1983b; Massey and Vandenbergh 1980, 1981; Coppola and Vandenbergh 1985). In microtines this effect is known from *Microtus pinetorum* (Schadler 1981) and *Pitymys subterraneus* (Frankiewicz and Marchlewska-Koj 1985). It is not known whether the peculiar mating system of *M. pinetorum* can be implicated, nor whether both species of pine voles exhibit similar habits. In mice, exposure to urine from grouped females stimulates isolated females to produce the chemical which inhibits puberty (Drickamer 1982). The intensity of inhibition is not dramatically affected by food deprivation or major changes to photoperiod (Drickamer 1984b), but there is some seasonality in the intensity (Drickamer 1984d). In direct contrast, urine from pregnant or lactating females accelerates puberty (Drickamer 1983b, 1984b), and urine from females in oestrus accelerates puberty in contrast to urine from nonoestrous females. Drickamer suggests these results reflect a system whereby mice indicate suitability of breeding conditions to their conspecifics, facilitating rapid growth of demes of mice when conditions permit. This view contrasts sharply with recent theory which interprets communication in terms of individual advantage (e.g. Dawkins and Krebs 1978; Caryl 1979; Wiley 1983; Krebs and Dawkins 1984; see Munn 1986 for a most amusing example), and should be viewed with some scepticism.

For microtines in the field, evidence of social suppression of sexual maturation by females is excellent for many species (Bujalska 1985a), but the physiological causes of this suppression cannot be currently inferred from laboratory studies. Many species of *Clethrionomys* exhibit retardation of maturation at high density (e.g. *C. glareolus* - Bujalska 1985a; *C. rufocanus* -Kalela 1957, Saitoh 1981; *C. rutilus*-Koshkina and Korotkov 1975), and maturation can be induced by experimental density reductions (Bujalska 1973; Saitoh 1981; Gilbert *et al.* 1986). In surprising contrast, monogamous *M. ochrogaster* show social suppression only at low densities (Getz and Hofmann 1986). Animals which successfully disperse from their natal nest always become reproductive. However, only three of 17 (18 per cent) females became reproductively activated while living in their natal nest. At high density, 37 of 48 (77 per cent) became activated. Because monogamy remains the modal mating system at high density, these data are inconsistent with a collapse in mating system causing a breakdown in social suppression. One possibility is that the suppression is mediated through the incest avoidance reported elsewhere, but the prospect of reproductive activation by a strange male increases as intrusions onto the home range occur at high densities. I have pointed out that incest avoidance in *M. ochrogaster* appears to be very different from that in other microtines (Table 3.4). Because the avoidance relies on failure of siblings to exhibit nasogenital grooming, rather than on lowered fertility in matings between sibling pairings, it may be easily disturbed by intrusion of a strange male.

Male effects on females.
Urine from a strange male accelerates puberty in young female house mice (Vandenbergh 1969), and in sufficiently many species of microtines to suggest that the effect is a general one (Richmond and Conaway 1969; Clarke and Clulow 1973; Hasler and Nalbandov 1974; Hasler and Banks 1975; Richmond and Stehn 1976; Baddaloo and Clulow 1981; Frankiewicz and Marchlewska-Koj 1985). In mice, unlike the interactions between females, grouping the males has no effect on the intensity of the result (Drickamer 1983a). However, dominant males produce a stronger effect than their subordinates (Drickamer 1983a). There are also seasonal effects (Drickamer 1984d).

One area which has attracted attention is the effect of kinship on puberty acceleration. Drickamer (1976) reported that female mice in litters with more than one male show delayed puberty in contrast to females with one or no brothers. However, his later work was less supportive of kinship effects (Drickamer 1983a, 1984a). More recently, Lendrem (1985) showed that while females exposed to the bedding of first cousins and unrelated males showed puberty acceleration, there was no effect from the bedding of fathers and uncles (Figure 5.3). The result for uncles is particularly important, as it implies some generalisation of kin recognition (Chapter 3). Fathers and uncles cannot be distinguished, but brothers and first cousins can.

As in all these analyses, information on microtine species, and wild popula-

GROWTH AND PUBERTY 115

Figure 5.3: Median date to first oestrus in female house mice exposed to soiled bedding from males of various degrees of relatedness. Vertical lines show standard error about the median. Females exposed to soiled bedding from fathers or uncles showed no significant acceleration of puberty, but bedding from cousins and unrelated animals exhibit accelerated puberty. After Lendrem (1985)

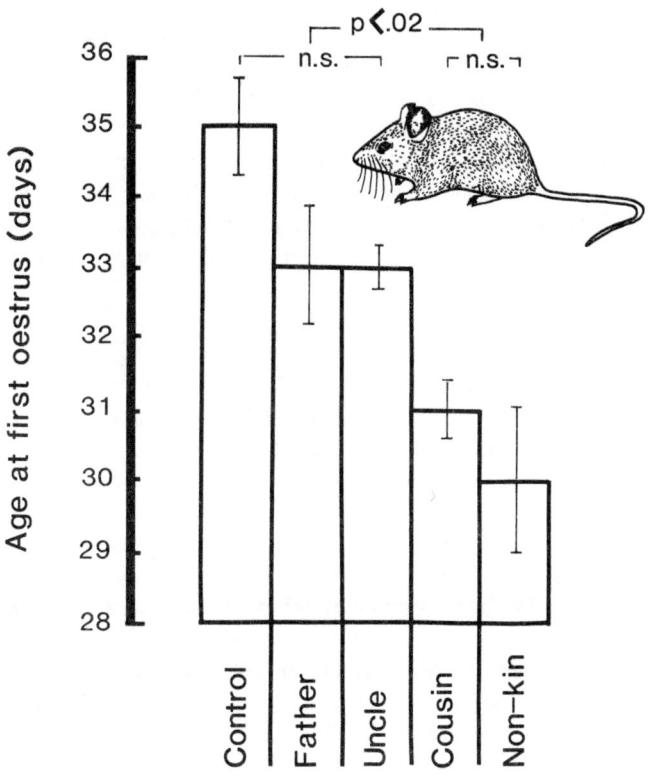

tions of rodents generally, are still far from satisfactory. Coopersmith and Banks (1983) provide the intriguing observation that sexual behaviour between pairs of lemmings (*Lemmus trimucronatus*) is facilitated by prior exposure of the female to the odour of the male. These results indicate that prolonged exposure may be important for full sexual facilitation in the field, and also support the idea that rodents may have a highly refined capacity to distinguish individual identity using odour (see also Huck and Banks 1979). I have already pointed out that sibling pairs of most microtine species do not reproduce (Table 3.4). This is either modulated directly through the failure of behavioural interactions to induce oestrus in the female (Carter et al. 1980), or by decreased fertility in females given the opportunity to mate with their brothers (e.g. Boyd and Blaustein 1985). This pattern is absent in *Microtus oeconomus* and *M.*

pennsylvanicus, for reasons which remain weakly understood, as Facemire and Batzli's (1983) claim that the mating system of these species is similar is probably untrue (Table 4.1).

In addition to these strong effects on juveniles, the presence of a male has been known for some time to shorten the oestrous cycle of adult females (the 'Whitten effect'), and the ability of strange males to induce pregnancy blockage in females is also well established (the 'Bruce effect').

Female effects on males.
The presence of a female appears to accelerate male sexual development in several rodents (Vandenbergh 1971; Maruniak, Coquelin and Bronson 1978), but suppress testicular function in rabbits (Bell and Mitchell 1984). Evidence for microtines is unsatisfactory, though chemical cues from the mother act more strongly than those of other family members in delaying puberty in *Microtus californicus* (Rissman, Sheffield, Kretzmann, Fortune and Johnston 1984).

Male effects on males.
Adult males suppress the sexual maturation of young males in a number of rodents, including microtines (Lecyk 1967; Lawton and Whitsett 1979; Vandenbergh 1980; see Gipps 1985 for a thorough review). In an intriguing analysis, Gustafsson, Andersson and Nyholm (1983) showed that there was geographical variation in susceptibility to social suppression of reproduction in male *Clethrionomys glareolus*. Males from northern Fennoscandian populations where populations fluctuated dramatically showed social suppression when housed *in groups*, but there was no effect in southern localities where population levels are much more stable (see also Gustafsson and Andersson 1980). In complete contrast, maturation of young males from non-cyclic populations in southern Sweden is suppressed when they are exposed to unfamiliar mature males, but males from cyclic populations are unaffected or stimulated (Figure 5.4). Gustafsson (1985) argues that these differences are likely to be genetically determined, but Hansson (1985b) claims that there are large differences between animals taken from the field and their F1 offspring, and favours a nutritional hypothesis for the difference (Hansson 1984c, 1985b; Hansson and Hentonnen 1985b).

Seasonal Generations

Changes in patterns of survival, growth and reproduction according to the season of birth are a conspicuous component of the biology of most small rodents living in seasonal environments. Animals born in the spring often grow rapidly and breed in the season of their birth. By contrast, animals born in the autumn attain weight only slowly, and do not breed until the following year. These differences have been called seasonal generations. Myllymäki's (1977a) analysis of demography in *Microtus agrestis* can be taken as representative. There is not only variation in growth between cohorts (Figure 5.5), but

Figure 5.4: Effect of exposure to conspecific males on testis development of male Clethrionomys glareolus *originating from cyclic (north Sweden) and non-cyclic (south Sweden) populations. (a) Relative weight of testis (testis weight × 10^3/body weight) in 60 day old male voles housed (1) alone, (2) with one other male, or (4) with three other males. (b) Relative weight of testis in 35 day old males exposed to (U) an unfamiliar male or (F) a familiar male. Data from Gustafsson (1985)*

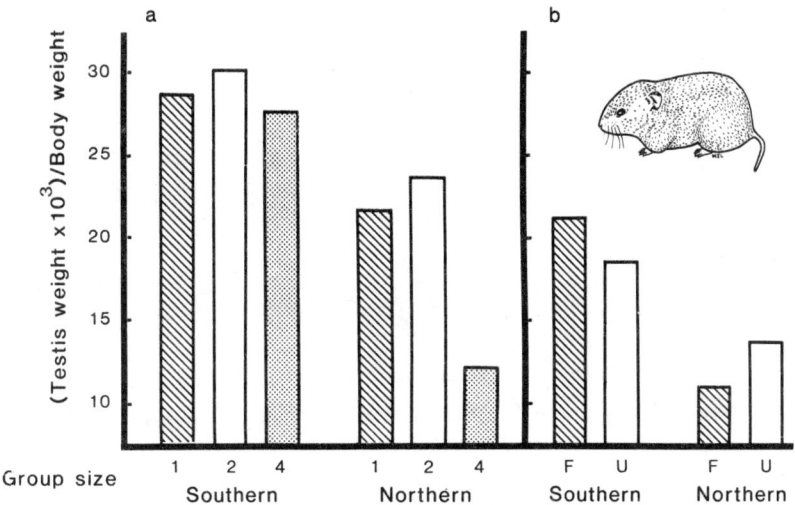

substantial individual variation within cohorts (Figure 5.6), which suggests that individual voles follow one of two growth trajectories. The fast growing spring generation almost never survive the winter. The data for other fluctuating populations of vertebrates are probably not of relevance here. Breeding in the season of birth in snowshoe hares is confined to the early increase phase (Vaughan and Keith 1980), and does not occur in grouse and ptarmigan. However, there is some evidence that blue grouse yearlings do not breed, apparently as a consequence of low adult mortality and its corollary of few vacant breeding sites (Hannon, Sopuck and Zwickel 1982).

The demographic changes in microtines are accompanied by numerous physiological changes which affect growth (Klevezal, Pucek and Malafeeva 1984), allocation of resources to particular organs (Yaskin 1984), and disposition of metabolically significant trace elements (Malzahn 1981, 1983). Many of the changes attributed to seasonal effects within individuals (Quay 1984) may warrant reinterpretation as individual differences in animals according to their season of birth (see also Mallory, Brooks and Elliott 1986). The combination of physiological and developmental changes are profound, and between season differences within a population may greatly exceed interpopulation and interspecific differences when these are expressed as means. I believe that understanding their ultimate and proximate causes must underlie any understanding of

118 GROWTH AND PUBERTY

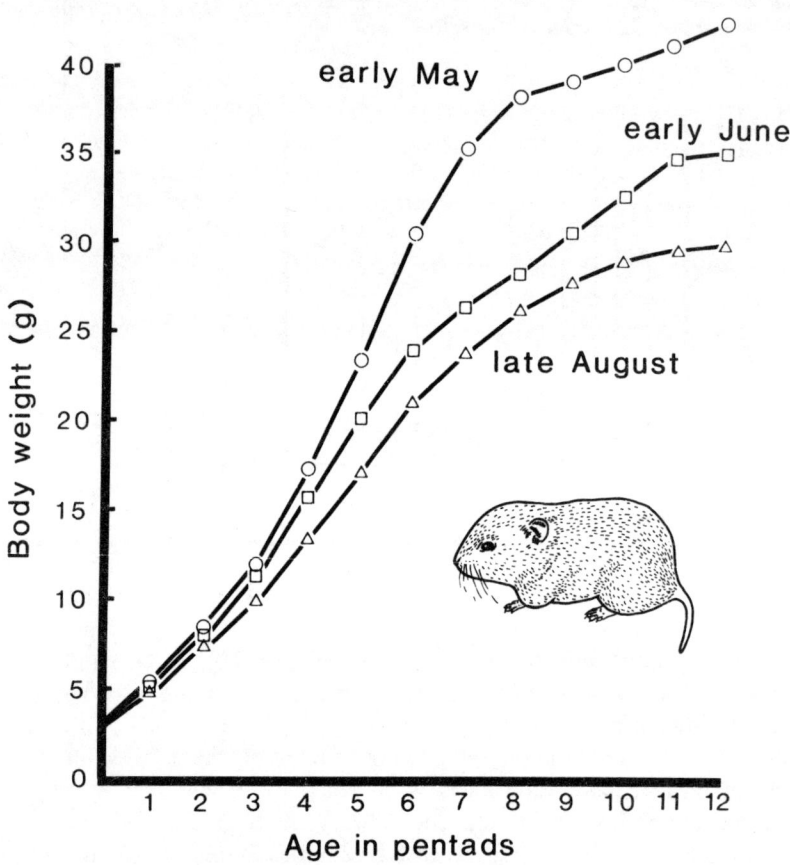

Figure 5.5: Growth patterns of cohorts of female Microtus agrestis *born at various times in the breeding season. (A pentad is five days.) Females born during May and June reproduced in the year of their birth. After Myllymäki (1977a)*

life history evolution in the microtines. The following discussion will illustrate just how elusive that understanding has proved to be.

Because population density increases during the breeding season, there have been many attempts to attribute the phenomenon of seasonal generations to social mediation of reproduction (Batzli *et al.* 1977; Lidicker 1979; Christian 1980; Lombardi and Terman 1980; Rissman *et al.* 1984). Experimental analyses provide some support for this hypothesis. For example, at peak densities almost no *Clethrionomys* mature in the year of their birth. Reduction in the number of adult females at this time enhances maturation in all species which have thus far been examined (Bujalska 1973; Saitoh 1981; Bondrup-Nielsen 1986; Gilbert *et al.* 1986).

Figure 5.6: Variation in body size and growth rates of different components of a cohort of Microtus agrestis *born during early June. After Myllymäki (1977a)*

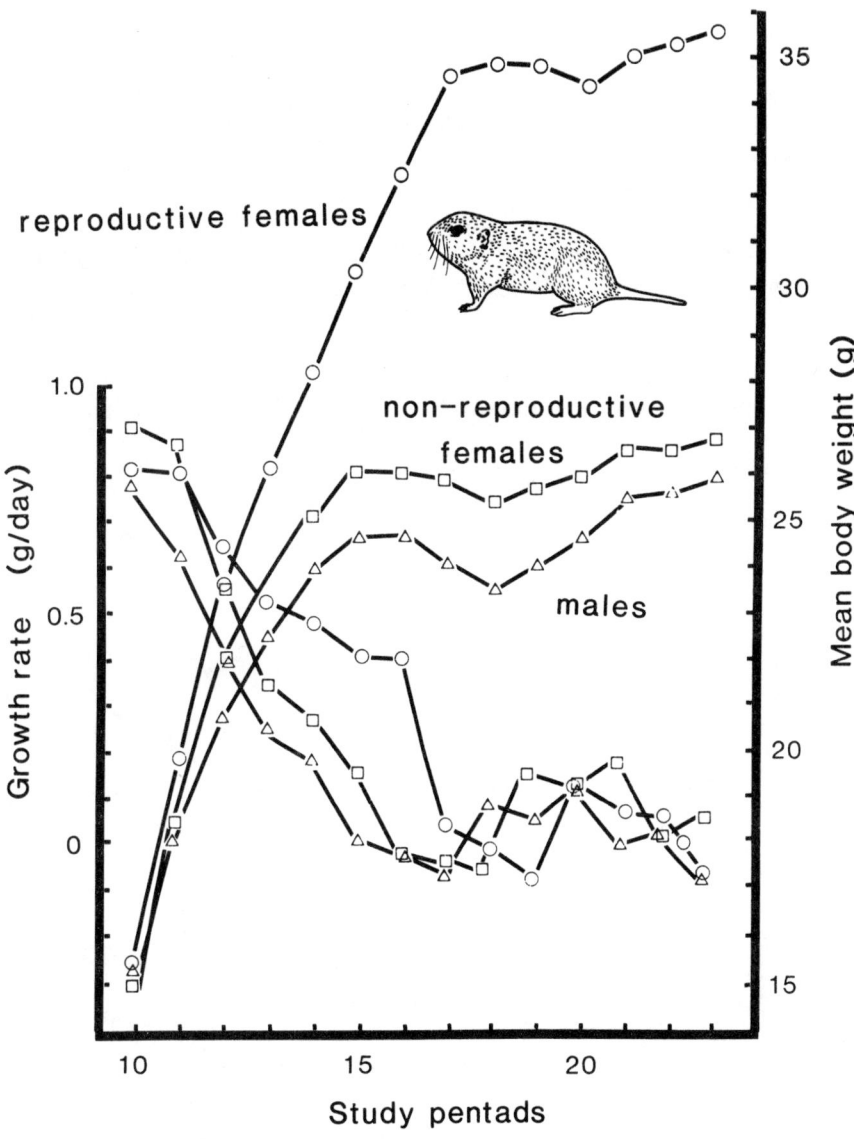

However, other studies suggest that interpretation of spring generations may involve considerable complexity. Photoperiodic influences on growth and development are well established for *Microtus montanus* (Pinter 1968; Petterborg 1978; Horton 1984), though nutrition is also implicated in stimulation of growth and initiation of reproduction (Berger *et al.* 1981; Sanders *et al.* 1981). Other microtines show less response to photoperiod (e.g. *Dicrostonyx groenlandicus*- Hasler, Buhl and Banks 1976). Insufficient data are available to attempt correlations between habitat and photoperiodic sensitivity. In a review of seasonal reproductive cycles in rodents, Zucker, Johnston and Frost (1980) make the point that research in biochronometry may be invaluable in understanding latitudinal differences in reproduction, and hence in understanding population dynamics. Quay (1980, 1984) has shown that the relative size of the pineal gland increases exponentially with latitude in the Rodentia (Figure 5.7), suggesting that there may be more pronounced sensitivity to photoperiod at high latitudes. Quay's data are confounded because of the limited latitudinal distribution of the families from which he draws data. Under these circumstances, it is difficult to distinguish between phylogenetic and adaptive

Figure 5.7: The relation between the relative volume of the pineal gland and latitudinal centre of distribution in rodent species. Open triangles = Sciuridae; Filled triangles = Muridae; Open squares = Gliridae; Filled squares = Capromyidae; Open diamonds = Bathyergidae; Filled diamonds = Caviidae; Open circles = non-microtine Cricetidae; Closed circles = Microtinae. Modified from Quay (1984)

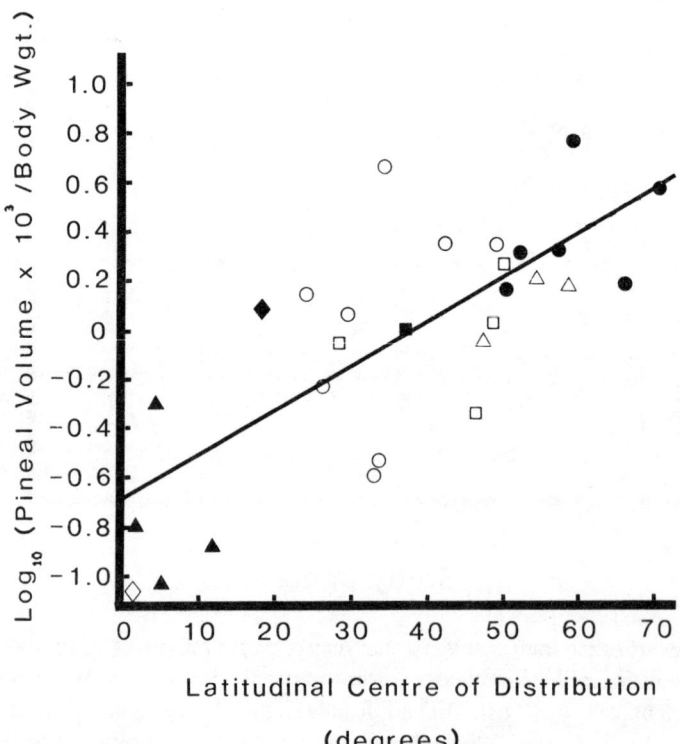

influences on the trait in question. A similar analysis which focuses exclusively on the microtines would be most illuminating.

Other authors have drawn attention to the different processes of physiological ageing in spring and autumn generations (Shvarts et al. 1964). While the former grows rapidly, most spring-born individuals are very short-lived. The first explicit gerontological analysis of natural populations was provided by Dapson, Feldman and Pane (1980), who examined *Peromyscus polionotus* in two different habitats. Chronological age could be determined by assaying the water-insoluble content of the eye lens. In the poorer habitat, reproductive effort and adult mortality were high in contrast to better habitat. Physiological ageing was defined as the accumulation of lipofuscin in Purkinje cells in the cerebellum, and occurred at a higher rate in the poorer conditions. The uncoupling of chronological and physiological age is probably a consequence of the stress associated with high rates of reproduction. Dapson (1979) showed that spring and autumn cohorts in the good habitat showed few differences in reproductive strategy, and most animals lived nearly two years. However, in the poor habitat, spring, summer and autumn born cohorts differed markedly in reproductive tactics. The cohorts born late in the season exhibited moderate reproduction at a level which did not increase the risk of mortality. Intense reproduction (e.g. simultaneous lactation and pregnancy) was most pronounced when the prospects of future reproduction were negligible (Table 5.5). These data suggest that the decline in reproduction by autumn cohorts may reflect optimisation of lifetime reproductive performance, and that the appearance of spring cohorts is most probable where lifespan is less than or approaches one year, so that the probability of breeding over two seasons is very low. Dapson's data also draw attention to the role of habitat heterogeneity in determining life history performance and patterns of growth.

In a related observation, Cockburn and Lidicker (1983) showed that *Microtus californicus* continued to breed in rapidly drying habitat in which they had no

Table 5.5: Influence of reproductive performance and time of birth on early autumn survival in *Peromyscus polionotus* living in suboptimal habitat. Double litters are the result of post-partum estrus. Females born early in the year survive very poorly during the autumn, but non-breeding animals survive well. Data from Dapson (1979)

Maternal birth month	Reproductive performance	Percentage surviving	
		September	November
Before June	6% non-breeding 437% single litters 47% double litters	17	2
June-July	50% non-breeding 50% single litters	11	6
September	100% non-breeding	45	36

prospect of surviving the summer, but entered reproductive quiescence wherever they had access to some green vegetation. Because the spatial scale examined by Cockburn and Lidicker (1983) was extremely small, it appears clear that analysis of reproductive prospects should pay attention to both macro- and micro-habitat heterogeneity. Growth may also alter in a manner consistent with adaptive interpretation. Gyug and Milar (1981) showed that although seasonal generations of *Peromyscus* spp. grew in a similar fashion until 40 days of age, subsequent growth depended on the prospects for future reproduction. Young mice with the potential to mature in the summer of their birth grew to adult size, but those which could not stopped growing when 75 to 80 per cent of adult size was reached. Because loss of body weight is a common characteristic in small mammal populations during winter (Stenseth 1978c; Ure 1984), it is easy to imagine that such differences in growth represent alternative tactics, with the spring generation maximising early reproduction and the autumn generation minimising the risk of failing to survive the winter.

The remaining set of hypotheses concerning the development of seasonal generations implicate selection on a very short time scale. House mice on the island of Skokholm show clear genetic and morphological differences according to the season of birth. Berry (1970, 1977, 1981) has argued that these differences suggest the operation of endocyclic selection for individuals capable of reproducing well, and those capable of surviving a population crash during the winter (Figure 5.8). Winter imposes heavy mortality, and selects for individuals

Figure 5.8: Endocyclical seasonal selection in feral house mice on the island of Skokholm. Modified from Berry (1970)

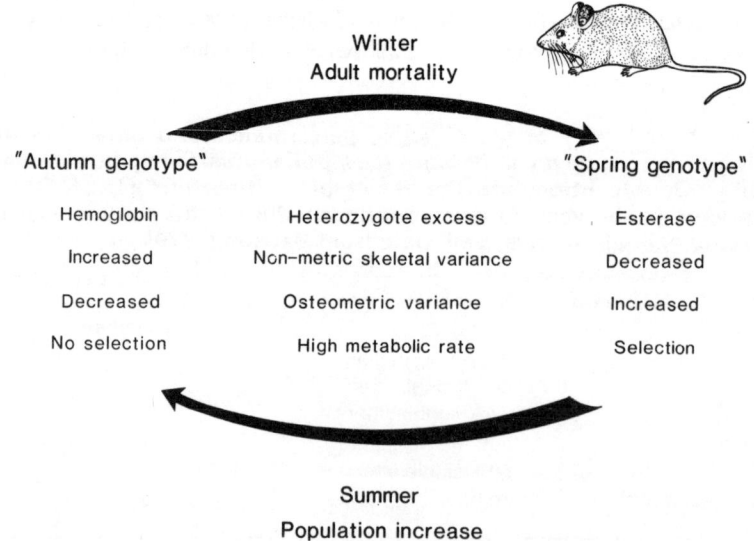

with a high metabolic rate. Osteometric characters show evidence of destabilising selection, while non-metric skeletal variance decreases. By contrast, during the breeding season there is apparently random survival of physiological variants. Electrophoretic evidence provided the best support for a genetic basis for these changes. For example, frequency of heterozygotes at the haemoglobin (*Hbb*) locus increased during the breeding season, and declined during the winter. Because the result is repeatable on several other islands where winter conditions are very severe, but not in tropical latitudes (see Berry 1981 for a review), selection rather than drift appears to cause the changes in allele frequency. Berry (1981) interprets these patterns in terms of antagonistic pleiotropy in the effects of gene products early and later in life. There is certainly very high heritability for age of maturity in mice (Drickamer 1981a, 1981b, 1983c), and one means of maintaining high additive genetic variance in natural populations is through antagonistic pleiotropy (Falconer 1981; Rose 1985). Gerontological studies of *Drosophila* support a role for pleiotropy (Rose and Charlesworth 1980; Rose 1984a), but the significance in mouse populations requires further examination.

Comparable data for microtine populations are being reported with increasing frequency (e.g. in *Microtus californicus* - Gill 1977; *Clethrionomys glareolus* - Fedyk and Gebcynski 1980; *C. gapperi* - Mihok, Fuller, Canham and McPhee 1983). Mihok *et al.* (1983) observed a repeatable increase in the frequency of a transferrin allele (Tf^p) in populations declining during the winter from high non-breeding density in the autumn. These authors have suggested that patterns of this sort may underly correlations between allele frequencies and population density. A high incidence of transferrin heterozygotes occurred in young voles that matured in the year of their birth. Anderson (1975) reported high heritabilities for juvenile growth rate and age at sexual maturity in *Microtus townsendii*.

The evolutionary implications of this argument are in complete contrast to the view that seasonal generations represent an adaptive response to seasonal conditions of plenty or adversity. Instead, selection throughout the non-reproductive and reproductive seasons breaks down adaptation for the forthcoming season. For example, in Skokholm mice, animals breeding in the spring may not benefit from a high metabolic rate, but selection during the winter means that the characteristics of the reproductives will be shaped by the stringency of the non-reproductive season. Relaxation of selection during the breeding season may lead to mass mortality during winter. Both these evolutionary hypotheses are quite distinct from the hypothesis that a decline in the quality of food or overcrowding are principal contributors to the suppression of development in the autumn generation.

Distinction between these alternative hypotheses is extremely difficult with available data. The only experimental effort attempts to test the social suppression hypothesis, and gives contradictory results. The descriptive analyses are convincing and difficult to criticise. If anything, everything works, lending support to the idea that seasonal generations and dynamics are crucial to an understanding of microtine behaviour and reproductive tactics. Unfortunately the winter ecology of small mammal populations has been much neglected, and syntheses of information are only just emerging (Merritt 1984). Although I feel that it would be premature to draw any conclusions about this phenomenon, it may be possible at this stage to design an experimental protocol which facilitates investigation of these hypotheses. The hypotheses and their major predictions are shown in Table 5.6.

It is also worth mentioning in passing that elements of each of the major hypotheses concerning population regulation in microtines is echoed in the development of hypotheses to account for seasonal generations. For example, rapid selection is an essential component of the Chitty hypothesis, social modulation of reproduction critical to the Christian hypothesis, and a decline in

Table 5.6: Hypothesised causes of spring generations, and experiments or descriptive studies which may help distinguish between them. It is important to recognise that the hypotheses are not as exclusive as their evolutionary causes and consequences. For example, it is conceivable that predictable declines in nutrition could select for photoperiodic modulation of growth (Pinter 1968; Martinet and Spitz 1971)

Hypotheses which suggest that the autumn generation is maladaptive

1. Manipulation by, and social interaction with, older animals depresses the growth and suppresses the reproduction of juveniles to minimise competition
    — keep density of breeding animals constant by repeated removal. The autumn generation should not differ from the spring generation.
2. Declining nutritional quality of habitat depresses growth (e.g. Hyvarinen and Heikura 1971; Hansson 1972)
    — There should not be spring generations in laboratory colonies supplied *ad libitum* food
3. Selection for high fecundity, and absence of selection for high rates of metabolism during the breeding season, depresses winter survival (e.g. Berry 1970)
    — Use quantitative genetic analyses to determine heritabilities of the characters of importance, and the extent to which antagonistic pleiotropy constrains their selection.

Hypotheses which suggest that both generations are well adapted

1. Photoperiodic or social cues cause optimisation of reproduction and growth relative to residual reproductive value (e.g. Dapson 1979)
    — Demonstrate photoperiodic sensitivity
    — Examine correlations between RRV and reproductive effort
2. Growth is suppressed to enhance winter survival, at the expense of reproductive capacity in the spring (e.g. Innes and Millar 1981).
    — Measure survival of cohorts of different body size when released to a common winter environment
    — Measure reproductive performance of individuals of differing growth patterns which have survived the winter.

food availability implicated in most extrinsic hypotheses of population regulation.

Development and Selection: a Quantitative Approach?

The observations that the pattern of growth and onset of reproduction are highly variable within microtine populations suggests that consideration of ontogeny is crucial in an analysis of life history evolution. Development has also been the focus of quantitative genetical analyses in vertebrates, and is a useful starting point to develop this topic further.

In order to illustrate the complexity inherent in quantitative genetical analyses of life histories, it is useful to consider a number of empirical examples. These examples are clearly relevant to the problems I have identified with the interpretation of life histories of microtines living in a seasonal environment. Berven and Gill (1983) examined heritabilities of larval developmental time and larval body size of *Rana sylvatica*, a frog with a broad distribution in eastern North America. Frogs from three sites which differed in latitude and altitude were studied (Table 5.7). In tundra sites shortened developmental time is strongly selected, presumably because of the short time available for larval development and metamorphosis; the length of the larval period is strongly canalised in contrast to larval body size, where some heritability is retained. In more equable lowland sites in Maryland, large animals compete more effectively. Body size consequently exhibits very low levels of heritability as a result of canalisation. In high altitude populations in Virginia, both traits are apparently important and have been canalised to an equal extent. Gene effects on the two traits are negatively correlated, presumably as a consequence of antagonistic pleiotropy. This negative correlation sustains high levels of genetic variability, and therefore there is potential for rapid response to selection, as additive genetic variance (and heritabilities) are necessary for the rapid response implied by seasonal selection models. Berven and Gill (1983) point out that these results have general implications for interpretation of the evolution of any

Table 5.7: Geographic variation in larval development traits of the wood frog (*Rana sylvatica*), and the heritability and genetic correlations between those traits. (From Berven and Gill 1983.)

Site	Lowlands	Mountains	Tundra
Environmental sensitivity			
Length of larval period	High	Medium	Low
Larval body size	Low	Medium	High
Heritability			
Length of larvel period	0.27	0.34	0.07
Larval body size	0.08	0.58	0.27
Genetic correlations			
Dev. rate vs. body size	0.65	−0.86	+0.09

life history trait where optimal expression is determined by a correlated response in another trait. It is probable that these include age and size at maturity, the number and size of offspring, and photoperiod response and developmental rates.

Strong seasonal selection on aspects of life history where biological interpretation is clear is not well documented in other higher vertebrates. Studies on the effects of seasonality, or periods of adversity, on small finches is a conspicuous exception. Bumpus (1899) provided one of the first examples of short-term selection when he observed mortality of small house sparrows under unusually poor weather conditions. More recently, Fleischer and Johnston (1984) have suggested that the relations between winter climate and selection are complex. They suggest that selection during winter favours increased body size in males, because it promotes dominance. However, selection favours reduction in size in females because of metabolic savings. Because male and female attributes are likely to be genetically correlated (Lande 1981), the outcomes of selection are likely to be complicated. Another source of genetic correlation may develop through antagonistic pleiotropy where there are differences between selection on adult and juvenile stages of life. Boag and Grant (1981) demonstrated intense selection for body size in a species of Darwin's finch (*Geospiza fortis*) during a severe drought. Large adults survived the drought much better than smaller animals, as their size facilitated handling of the foods left in their habitat. The intensity of this selection was partly a consequence of extremely high values of heritability for adult body size (> 0.8). The selection recurs under conditions of adversity (Price, Grant, Gibbs and Boag 1984). These results raised the question of whether selection for large body size is balanced by selection for small size at other stages of the life cycle. Price and Grant (1984) provided evidence that this was indeed the case, presenting data which suggest weak selection for small body size at two stages of development to adulthood, possibly because large juveniles suffer a metabolic disadvantage.

The alternative to genetic canalisation of traits is phenotypic plasticity. The second hypothesis for rodent seasonal generations which interprets both the behaviour of spring *and* autumn generations as adaptive, necessarily implies adaptive plasticity. Caswell (1983) suggests that plasticity will be adaptive under many conditions, but points to the weakness of current theory in treating a number of questions. Most notably, there are several ways in which plasticity may manifest itself. Smith-Gill (1983) has elaborated the extreme forms of developmental plasticity most clearly (Table 5.8). She distinguishes between:

(1) **developmental conversion**, in which organisms use specific environmental cues to alter genetic programmes of development. This may be brought about by either an on-off switch, which halts or commences development, or expression of alternative developmental programmes leading to discrete phenotypes; and

(2) environmentally induced **phenotypic modulation**, where phenotypic variation in development results from the relative sensitivity of various parts of the organism to primarily physical factors in the environment. The genetic

Table 5.8: Comparison of two types of developmental plasticity (from Smith-Gill 1983)

Developmental conversion	Phenotypic modulation
Specific genetic basis	No specific genetic basis; reflect physiology of the species
Specific response to specific set of environmental stimuli	Non-specific variation in developmental phenotype due to physiological constraints of the environment
Change in developmental programme	No change
Occurs at discrete developmental stages	At any time in life history; phenotypic results may be stage specific
Produces discrete phenotypes	Continuous phenotypic variation
Adaptive	Not necessarily adaptive; may be non-adaptive

programme is not altered, but rates and degrees of expression of the developmental programme may be modified. Phenotypic modulation need *not* be adaptive. Indeed, departures from the 'genetically optimum' developmental process have been the subject of some detailed analyses. Compensatory growth represents one adjustment to temporary deviations from a genetic programme, and even this phenomenon has substantial additive genetic components (Atchley 1984).

In contrast to the view that plasticity is important, Gillespie (1974, 1975, 1977) has argued that selection for homeostasis and reduction in variance in the face of environmental variability may be more important than selection for maximisation of reproductive output (see also Rubenstein 1982). Although the generality of this result is confounded in a number of ways (Bulmer 1984; Orzack 1985), the measurement and interpretation of variance are of profound importance for understanding life history attributes like growth and offspring number.

Adaptive seasonal variation in animal reproduction and non-reproductive diapause is well documented in many species of invertebrates and plants (e.g. Wellings, Leather and Dixon 1980; Hairston and Munns 1984). The response of growth to photoperiod in many species of microtines certainly supports the idea that in some populations developmental conversion may be implicated in growth and the acquisition of reproductive maturity. Stearns and Crandall (1981a) catalogue a variety of forces which could lead to adaptive delays in maturity for either adults or their young (Table 5.9). As these authors suggest, several of these forces may interact to promote evolution of delay in puberty.

Again, we are left with the paradox that theory supports a role for all the hypotheses implicated in the development of seasonal generations in rodents, and in particular supports a role for social suppression (phenotypic modulation),

Table 5.9: Hypotheses for the evolution of delays in maturity. The hypotheses are not mutually exclusive. Summarised from Stearns and Crandall (1981a)

(1) A reduction in the cost of reproduction and an increase in adult survival rates (e.g. Wittenberger 1979b)
(2) A gain in fecundity as a consequence of delay (e.g. Gadgil and Bossert 1970)
(3) A response to density-dependent resource shortage (e.g. Frogner 1980)
(4) Minimise the rate of decrease in a declining population (e.g. Mertz 1971)
(5) Enhancement of juvenile survival (e.g. Hirshfield and Tinkle 1977)
(6) Increased fitness in the face of unpredictable larval mortality. The delay is in the egg stage (e.g. Livdahl 1979)

endocyclical selection (with heritabilities maintained by antagonistic pleiotropy), and programmed changes to development (developmental conversion). This raises the prospect that there may be intraspecific and interspecific variation in the dominant mode of reproductive control, just as the potential for selection varies dramatically in Berven and Gill's (1983) wood frog study (see also Brooks and Webster 1984; Mallory et al. 1986). Such diversity would contribute to an understanding of the result that sensitivity to social suppression of reproduction varies with latitude, and that house mice on tropical islands show no evidence of the endocyclical selection reported from more extreme climates.

Predicting the circumstances under which each mode is likely to operate is a risky and speculative exercise, as no available data have been gathered with particular regard to the alternative hypotheses. However, endocyclical selection appears most likely where the potential for population expansion in the reproductive season is most extreme. This will depend on both the length of the breeding season in raltion to the generation time of the animal, and the nature of spatial patchiness in the environment (where colonisation and invasion habitats are extensive in contrast to constant habitat). This potential is likely to be low in some mammals living at extreme latitudes and altitudes, where the breeding season will be short, and habitat availability may be determined as much by succession as by seasonal factors (e.g. *Peromyscus* spp. - Millar, Wille and Iverson 1979; Millar 1982; Millar and Innes 1983; *Microtus xanthognathus* - Wolff and Lidicker 1980).

Developmental conversion may be favoured in environments where seasonality is pronounced but predictable, and photoperiodic sensitivity may well increase in the highly seasonal environments at high latitudes and altitudes. There is evidence that developmental conversion can be overridden by highly favourable nutritional conditions, which promote either winter breeding or early onset of breeding (Negus and Berger 1977; Hickie et al. 1982; Eriksson 1984; Hansson 1984b; Jannett 1984). These conditions should also cause relaxation of any endocyclical selection (Berry 1970). The role of social modulation within this framework requires further investigation at different densities. Whether high density acts as a trigger which delays development in a manner adaptive for the offspring, or represents adult manipulation of juvenile development is uncertain. Manipulation should be subject to counter adaptation by juveniles, particularly as physiological control is implicit in the suppression of puberty. It is also not generally known whether social suppression could occur at low densities but is ordinarily avoided by dispersal, or is a 'pathological' consequence of high densities, as is implicit in the Stress Hypothesis of Christian (1970). In either case it appears that the fluctuating social environment may have the capacity to override ordinary seasonal responses.

Summary and Prospectus

Among mammalian taxa, microtine rodents exhibit high rates of reproduction, even when dietary and allometric influences are taken into account. Despite this rapidity, there is evidence of substantial intraspecific variation. Although it is tempting to attribute much of this variation to density effects, appropriate paradigms for the study of density-dependence remain weakly developed.

Although metabolic measures can often be usefully used to infer tradeoffs between life history tactics, a more refined analysis should rely on measurement of the correlation structure of the traits themselves. The theory of quantitative genetics is well suited to analyses of this kind. The complexity of interaction is well illustrated by the variation in growth and puberty in small rodents living in seasonal environments. There are currently several hypotheses to account for the phenomenon of seasonal generations, or seasonal modulation of growth. Not only do current data fail to distinguish between the alternatives, but they also suggest that different causes may operate in different populations.

Topics for Future Research

(1) I believe that the phenomenon of seasonal generations underlies life history evolution in the microtines. Comparative data and explicit tests of the various hypotheses are required as a matter of urgency. If it is true that maladaptive seasonal selection with heritability maintained by antagonistic pleiotropy operates in some populations of small rodents, while adaptive variation in maturation schedules operates in others, microtines may reemerge like a Phoenix as a useful predictive tool for life history modelling in general.

(2) While I have made the point that we require a better understanding of both heritability and covariance of life history attributes many times, the data are of such importance that this need is worth reiterating.

Chapter 6
Life History Evolution: Reproductive Strategies

The conditions preceding maturity are clearly of profound importance in the interpretation of microtine life histories. However, the onset of reproduction is only a prelude to the iteroparous reproduction characteristic of most small mammals. In this Chapter, I examine the ways in which an animal may vary the allocation of reproductive effort, and then examine the interaction between life history tactics which dictate the strategy observed in microtines.

Because the methodological problems which plague other areas of research on secretive small mammals are less pronounced, the data on patterns of reproduction in the taxa of interest are rich and have been well treated on a number of occasions (e.g. Krebs and Myers 1974; Keller 1985). Rather than repeat reviews of this sort, I hope to focus mainly on aspects of reproductive performance which are relevant to the interpretation of social behaviour and population dynamics. In particular, I review some life history traits of unquestioned importance (lifespan, litter size and sex allocation), and then discuss the interaction between those traits and time to maturity. I place particular emphasis on the themes which recur throughout the literature on fluctuating populations, the presence of latitudinal gradients in cyclicity, in seasonality, and in the expression of life histories.

Lifespan

Most microtines live less than one year, breed in only one season, and the number of litters they produce is determined by the period habitat remains favourable for reproduction. Thus lifespan will vary more with time than with the length of time over which an animal may reproduce. For example, *Microtus xanthognathus* appear to produce only two litters in their lifetime (Wolff and Lidicker 1980), approaching semelparity more closely than any other eutherian (see Braithwaite and Lee 1979). Occasional anomalies have been reported. Rose and Dueser (1980) noted a high frequency (>5 per cent) of *Microtus pennsylvanicus* living more than twelve months in a population living in an equable environment in Virginia. There was no evidence of social suppression or delayed growth in this population, again drawing attention to the importance of seasonality as a driving force behind microtine life history evolution.

Litter Size

During the period an animal remains capable of producing young, there may be variation in the number and size of young, the type of young (if young are polymorphic, e.g. for sex), and whether the young are tolerated or forced to disperse. Because variation in the number of offspring produced can influence the rate of population growth, considerable attention has been paid to the factors which influence the litter size of microtine rodents.

Intraspecific variation is often dramatic. Substantial effects have been described to result from age, size, and parity in all of the vertebrates which fluctuate in abundance (Cary and Keith 1979; Hannon and Smith 1984; Keller 1985; Chapter 3; Figures 1.5, 5.2), but none of these effects are unique to cyclic populations (e.g. Perrins 1965; Clutton-Brock 1984). Interannual effects have been frequently reported, but these remain of uncertain importance, and difficult to dissect from causes unrelated to density (Krebs and Myers 1974). Seasonal effects are often pronounced, and might also be attributed to a variety of factors, as age structure, prior sexual experience and body size will usually vary during the course of a season.

Despite the difficulties in distinguishing these confounding effects, increasing evidence supports a role for nutrition. Litter size is strongly correlated with incident radiation in *Clethrionomys glareolus* and *Microtus arvalis* (Zejda and Pelikan 1984), suggesting a role for energy availability in determining litter size. Seasonal fluctuations have been shown to be unambiguously dependent on diet in *M. californicus* (Krohne 1980, 1981). Krohne compared litter size in *M. californicus* living in native perennial grassland and in grassland dominated by introduced annual pasture plants. In the perennial area, breeding occurred throughout the year, but litter size was consistently low except during a 'spring flush' when most food species were growing rapidly (Figure 6.1). Annual grassland supported reproduction over a much shorter period, but the litter size was consistently high. These results can be related to the nature of the grasses which

Figure 6.1: Litter size in two populations of Microtus californicus. *Vertical bars are standard errors. The population living in annual grassland does not breed in summer. The convergence of the two curves takes place during a spring flush of vegetation. (After Krohne 1980)*

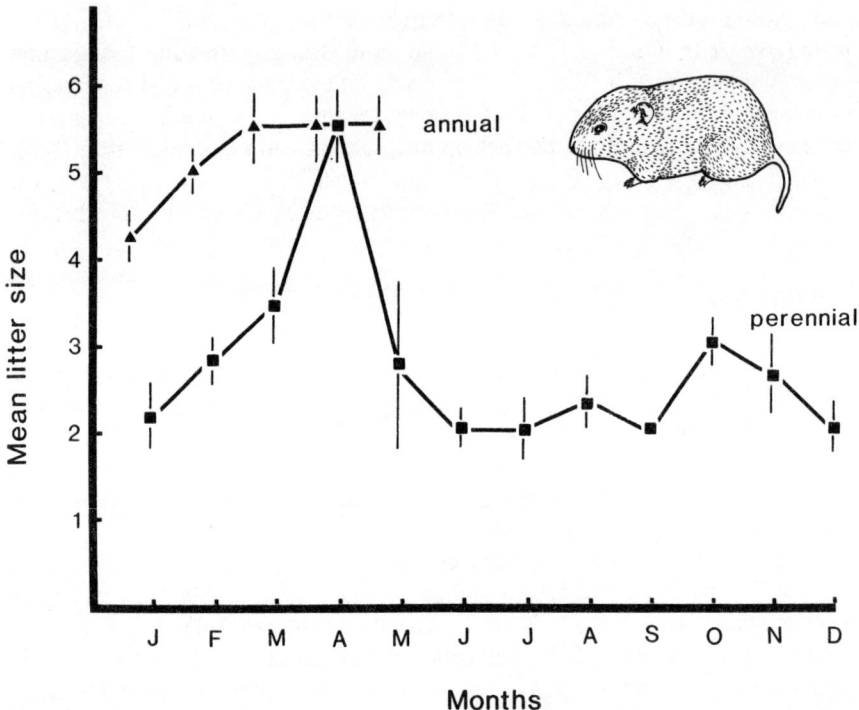

make up the diet of *M. californicus*. Perennials contain much more indigestible carbohydrate than annual species, and this fibrous material accelerates gut passage, reducing the amount of nutriment which can be extracted from the grass. During rapid growth, fibre is less abundant in the soft tissues of the rapidly growing plants, and carbohydrates will be more easily accessible. In the annual plants, growth in California is confined to the rainy season in spring and winter. Diet supplementation has increased litter size of *M. ochrogaster* (Cole and Batzli 1978).

Social modulation of reproduction has already been discussed, but it is worth reiterating that failure of incest avoidance and reproductive priming are likely to have a negative and positive effect on litter size, respectively. Reproductive debilitation under conditions of crowding or intense competition has been predicted to reduce litter size by Christian and his supporters (see Chapter 1). Unambiguous evidence for such an effect in natural populations is not available.

There are few reliable estimates of heritability of litter size in microtines, and such estimates will undoubtedly prove to be extremely difficult to isolate from maternal effects. Low heritability of litter size in domestic animals and laboratory rodents is well documented (Falconer 1981). Low correlations have been expected because the strong correlation between litter size and fitness should promote strong selection, and eliminate genetic variance. The few studies which are available for wild populations of higher vertebrates other than microtines report some evidence for higher heritabilities (Perrins and Jones 1973; van Noordwijk et al. 1981; Flux and Flux 1982; Findlay and Cooke 1983). As environmental and social factors are often of overwhelming importance (McCleery and Perrins 1985), the maintenance of these high heritabilities may be indicative of negative genetic correlations of the sort which appear to influence growth and maturity in house mice. Krohne (1981) calculated that heritability in laboratory colonies of *Microtus californicus* was in the vicinity of 17 per cent, but any effects were likely to be swamped by the habitat effects described above. There are differences in the sensitivity to parity effects of litter size in two genetically distinct colour morphs of *M. californicus* (Figure 6.2) (Gill 1976), and heterozygote advantage in crosses between another two color morphs (Gill 1977).

Latitudinal Trends

Stenseth, Gustafsson, Hansson and Ugland (1985) show that there is a gradient

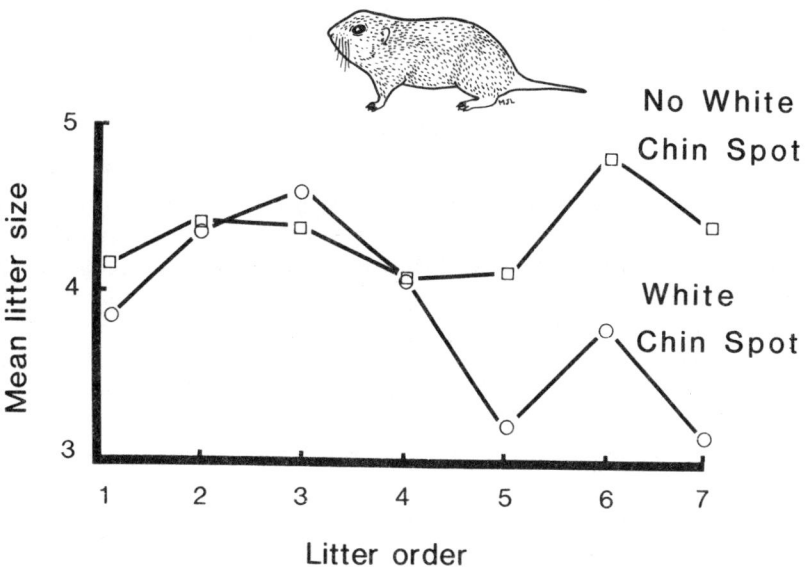

Figure 6.2: *The effect of litter order on litter size in two genotypes of* Microtus californicus. *(After Gill 1976)*

of litter size in *Microtus agrestis* and *Clethrionomys glareolus* associated with latitudinal changes in the extent of cyclicity in Fennoscandia (Table 5.2). These latitudinal differences have been used to support the idea that litter size is heritable (Gustafsson, Anderson and Westlin 1983; Gustafsson and Batzli 1985). However, other authors have had diffficulty repeating these results (Hansson 1984c; Hansson and Hentonnen 1985b).

Although Stenseth *et al.* (1985) attempt to relate this gradient to that of cyclicity, the tendency of brood size to increase with latitude has been documented for many bird and mammal species (Lack 1954; Lord 1960; Klomp 1970; Cockburn, Lee and Martin 1983), though the trend is by no means universal (e.g. Zammuto and Millar 1985). It seems premature to attribute the difference to cycling, when many alternative hypotheses are available.

Predation.
Cody (1966) has suggested that within a bout of reproduction, an animal has to optimise expenditure on maintenance, reproduction and other activities, particularly avoidance of predation. Any increased reproductive effort will be accompanied by costs associated with maintenance or increased risk of predation. I have already reviewed evidence that the gradient of cyclicity in microtines is associated with transition from exposure to generalist predators in non-cyclic populations to exposure to microtine specialists at extreme latitudes where cyclicity is most pronounced. Which of these predator types is most likely to increase the costs of predator defence, or whether defence of the nest is at all effective is unknown. In the populations exposed to specialist predators, there should be relaxation of the risk of predation in order to explain the latitudinal trend, but it would be stretching credibility to explain microtine litter sizes in this light.

Food availability.
Lack (1947, 1954) originally proposed that because diurnal birds had longer to feed during summer at extreme latitudes, they would have more food to devote to rearing offspring. This hypothesis was falsified by the replication of latitudinal gradients in nocturnal species of both birds (Owen 1977) and mammals (e.g. Lord 1960), and the demonstration that clutch size in some birds continues to increase beyond the Arctic Circle, where increments in day length can no longer occur (Hussell 1972). Ashmole (1963) provided a modified hypothesis which suggested that the relevant measure of food availability was the difference between availability at the time the population was regulated (usually the winter), and the abundance of food at the peak of reproductive effort (Figure 6.3). Brood size should increase with seasonality, accounting for both latitudinal and altitudinal trends (see also Hussell 1985). Qualified support for this hypothesis is available for birds (Ricklefs 1980; Koenig 1984) and dasyurid marsupials (Cockburn *et al.* 1983). There is good evidence that nutrition influences litter

Figure 6.3: The Ashmole hypothesis for variation in brood size. Because the adult population is often regulated during periods when resources are scarce, and animals breed when resources are abundant, the additional resources available to any animals which survive the non-breeding season will vary with the amplitude of seasonality. This amplitude will determine the resources which can be diverted to reproduction without risk, and will therefore determine the brood size. (After Ricklefs 1980)

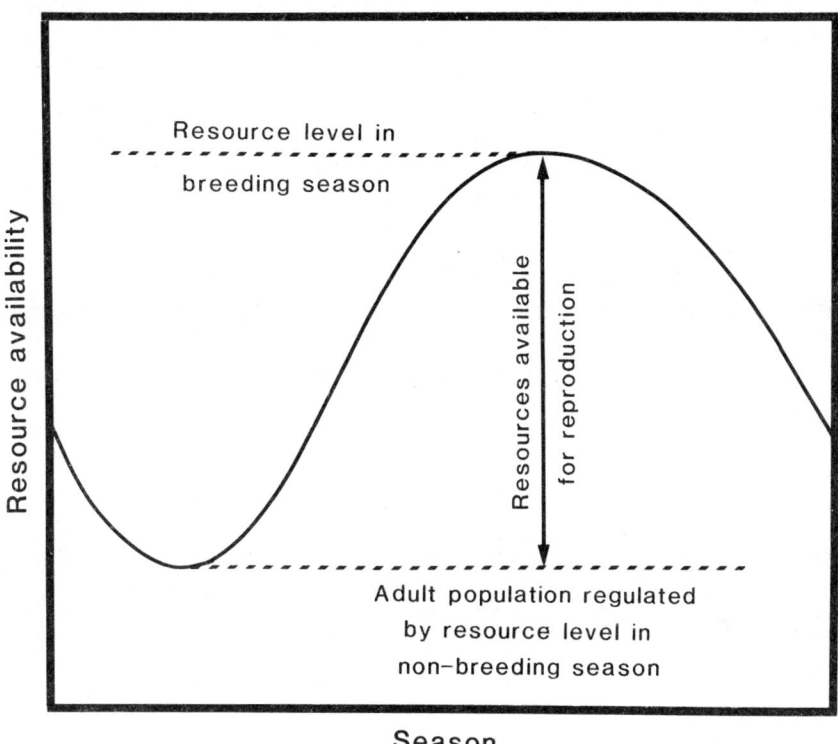

size in microtines, and population size almost always declines during the winter, or season of greatest stringency. This hypothesis may repay greater attention.

Tradeoff against Residual Reproductive Value.

Several authors have argued that breeding within the current bout should be deferred to maximise the probability of breeding again (Charnov and Krebs 1974). As the probability of living to breed twice is postulated to be higher in the tropics than at extreme latitudes, this hypothesis can be translated into an explanation of latitudinal trends. However in contrast to the central prediction of this hypothesis, latitudinal and altitudinal gradients have been demonstrated in a semelparous mammal, *Antechinus stuartii*, in which the prospects of any trade-off are precluded by the unusual life history (Cockburn *et al.* 1983). This does not mean that tradeoffs do not influence the evolution of litter size (see discussion of metabolic costs above), just that they may not explain latitudinal gradients.

Further, the widespread assumption that an increase in brood size commonly leads to reduction in survival has been questioned several times (Högstedt 1980; Smith 1981; Alerstam and Högstedt 1984; Richter 1984). If an animal has access to superior resources it may have higher fecundity and higher survival. In particular, there are likely to be strong individual differences in the extent to which costs are suffered (Lomnicki 1978, 1980, 1982; Begon 1984; Lomnicki and Orbach 1984; Sutherland and Parker 1985), and some animals may be completely insulated from any costs. Among microtines, there is abundant evidence of increased mortality associated with the onset of reproduction (e.g. *Microtus californicus* - Cockburn and Lidicker 1983), but no data pertaining to variance in these effects, except in response to variation in habitat characteristics.

Interspecific Differences

Because some species occupy limited geographical ranges, and there are marked variation in diet and social organisation, any of the hypotheses I have just discussed might also contribute to interspecific differences within the microtines. As for the intraspecific case, there is a good correlation between litter size and the changes in climate associated with increased latitude and altitude for both *Microtus* and *Clethrionomys* spp. (Innes 1978).

Microtus xanthognathus has the largest average litter size reported for any microtine species (8.8 young; Wolff and Lidicker 1980), and apparently has a very low probability of repeated reproduction. This would be consistent with intense reproduction because residual reproductive value is low. However, there is little evidence of metabolic debilitation associated with reproduction. Adult females often participate in nesting constellations throughout the winter (Wolff 1980), even when the prospect of future reproduction is minimal. Consistent with a nutritional hypothesis, the lowest litter sizes are reported for *Phenacomys longicaudus*, which feeds on conifer needles (Hamilton 1962). Data on reproductive contributions by microtines living in communal systems will be eagerly awaited, but will require very careful assessment of patterns of maternity and paternity.

One very interesting anomaly is the low litter sizes in the polyandrous and monogamous species, *Microtus pinetorum* and *M. ochrogaster* respectively. It has often been suggested that there is a general correlation between fecundity and obligate monogamy in mammals (Kleiman 1981). This is because a female capable of producing large numbers of young with male assistance should not need male assistance to rear at least some young, reducing the incentives for males to provide paternal care and increasing the potential for polygyny. Low litter size is less likely to be correlated with the facultative monogamy implicated in the resource dispersion hypothesis advanced by Getz (1978) and Getz and Carter (1980), which I discussed in Chapter 4.

Summary

Although there is good evidence for intraspecific and interspecific variation in litter size, the evidence for a genetic basis for this variation is extremely weak. Lack of evidence for genotypic control of variation calls into question theoretical models which implicitly assume this control. Nutritional hypotheses provide promising leads to both intraspecific and interspecific differences, but experimental evidence is almost non-existent. Indeed, Davies and Lundberg (1985) have recently pointed out that experiments which supplement the food available to passerine birds through the time of laying are much more likely to bring forward the time of laying than increase clutch size, particularly in small species.

Sex Allocation

One opportunity small mammals have to influence the type of offspring they produce is through control of sex allocation. However, the possibility of manipulation of the primary sex ratio in mammals is restricted by the simple Mendelian and chromosomal mechanism of sex determination, and the evidence for sex ratio control remains scanty (Williams 1979; Charnov 1982; Clutton-Brock and Albon 1982; Clutton-Brock 1985). This has led to the suggestion that it is easier to manipulate the extent of investment in each sex, rather than the relative number of male and female young (Maynard Smith 1980; Clutton-Brock, Albon and Guinness 1981).

One tempting and apparently repeatable observation which is suggestive of differential resource allocation to male and female young is the tendency of brood reduction to affect male embryos and neonates much more dramatically than female embryos and neonates. For example, McClure (1981) showed that food-restricted wood rats (*Neotoma floridana*) often failed to rear male offspring while survival of female offspring was less dramatically impaired. This may be one reason why male-biased secondary sex ratios are much more prevalent than female bias (Clutton-Brock and Albon 1982).

The most useful hypothesis for the interpretation of sex-biased brood reduction was proposed by Trivers and Willard (1973). They argue that in a polygynous mating system, males will have more variable reproductive success than females, because fewer males will gain access to females than vice versa (see Chapter 4). Males in better condition than other males will consequently obtain greater advantage than females having the same advantage of condition over other females. Thus females in poor condition should produce a predominance of female offspring, each with a moderate chance of reproductive success, while those in good condition should produce males, with the prospect of a large payoff in grandchildren. A number of laboratory studies of food-deprived small rodents have been interpreted as providing evidence for this hypothesis (Labov *et al.* 1985). In each instance, food-restricted mothers reduced the size of their litters, always at the expense of their male offspring (Rivers and Crawford 1974; McClure 1981; Labov, Huck, Vaswani and Lisk 1986). However, another interpretation of these data is possible. Clutton-Brock, Albon and Guinness (1985) point out that male offspring are more susceptible to hardship than female offspring, probably as a result of sexual selection for rapid growth. For an amusing and salutory review of the generality of this pattern, see Trivers (1985). For rodent data, see Widdowson (1976). Under these circumstances, it may be very difficult to use female-biased sex allocation to indicate support for the Trivers-Willard hypothesis, though such evidence is now available for red deer (Clutton-Brock, Albon and Guinness 1986). In any case, intrapopulation patterns are likely to be complex. Myers *et al.* (1985) have recently suggested that the short-term effects of weather alter sex-specific patterns of intrauterine survival of *Peromyscus maniculatus bairdii* in complex ways. Relating laboratory studies to the field will be extremely difficult, and dependent on much larger samples than are usually gathered (e.g. Gosling, Baker and Wright 1984).

A recent study of microtines in the wild offers a very promising lead. McShea and Madison (1986) showed that there were large differences in the sex ratio of juvenile voles recruited into a population of *Microtus pennsylvanicus*. Spring recruits were predominantly female, while autumn recruits were usually male. This pattern was not reflected in the sex ratio at birth, suggesting that differential mortality occurred between birth and first capture. Maternal influence was implicated by the observation that female pups were heavier than males in spring litters, and male pups heavier than females in autumn litters. Because the lightest member of the litter was most prone to mortality, differential investment may influence these results. McShea and Madison (1986) argued that their data were consistent with manipulation in response to the sex-specific probability of reproductive success. In particular, females should invest most heavily in the sex which will breed sooner. Because female *M. pennsylvanicus* are territorial, and females limit recruitment, the prospects of successful recruitment and reproduction of juvenile females will be density-dependent (Boonstra and Rodd 1983). Early in the breeding season when densities are low, recruitment of females should be possible, but sons will have to compete against large males which have successfully overwintered. Greater investment in daughters may therefore be favoured. By autumn, breeding spaces for females will be saturated but males may still have the opportunity to breed because they do not compete with territorial females for space, promoting investment in sons. McShea and Madison (1986) determined the sex ratio pre-partum by autopsy and found a weight-dependent effect (Table 6.1). Heavy females have few sons in autumn, contrary to the conclusions drawn concerning post-partum allocation. The significance of this effect is uncertain, but may be explicable as terminal investment by females which are unlikely to compete with their daughters for future reproductive opportunities (see also Cockburn, Scott and Dickman 1985). It is most intriguing that alteration of investment of this sort should occur in a

Table 6.1: The sex ratio (per cent males) of pups produced by *Microtus pennsylvanicus* in relation to season and maternal body weight. Median weight was 43 grams. Sample size is in parentheses. The sex ratio of autumn litters is significantly affected by maternal weight. Data from McShea and Madison (1986)

	Light females ($\leq$43 g)	Heavy females ($>$43 g)
Spring	36(131)	40(215)
Autumn	52(139)	24(119)

microtine species where social suppression of maturation in response to the presence of siblings is poorly developed (Table 3.4; Batzli, Getz and Hurley 1977; Porter and Dueser 1986). Variation in sex allocation may render incest avoidance unnecessary. Although their data were inadequate to draw reliable conclusions about sex allocation, Anderson, Whitney and Huang (1976) also reported a strong bias towards daughters among early spring recruits into a population of *M. richardsoni*, which appears to have a mating system similar to that of *M. pennsylvanicus* (Ludwig 1984). By contrast, Berry and Jakobson (1971) report an opposite pattern in an island population of the house mouse. These relations requires further investigation in both the laboratory and the field.

In a related analysis, Clark, Spencer and Galef (1986) demonstrated bimodality in sexual maturation in female gerbils (*Meriones unguiculatus*). Some gerbils exhibit vaginal perforation before eye-opening; the remainder do so after weaning. The relative frequency of each type exhibits a circannual rhythm, even in constant laboratory conditions. The differences in maturity are reflected in numerous differences in the pattern of reproductive investment (Table 6.2). Importantly, early maturing females produce a predominance of daughters at both birth and weaning, and the type of daughter born in each litter reflects maternal phenotype. Clark *et al.* (1986) are reluctant to propose adaptive interpretation of this pattern of sex allocation, but draw attention to Drickamer's (1976) observations relating sex ratio of offspring to the pattern of sex allocation (see above) and the prediction that sex differences in intergeneration overlap may lead to seasonal shifts in sex allocation (Werren and Charnov 1978; see also Seger 1983; Werren and Taylor 1984; Cockburn, Scott and Dickman 1985).

The Unusual Case of the Lemmings

It does appear to be generally true that mammals will have little opportunity to influence sex allocation, as the sexes can be distinguished by homomorphism (XX females) or heteromorphism (XY males) in one pair of chromosomes (Williams 1979). However, the best exception to this general rule exists in two genera of lemmings (*Myopus* and *Dicrostonyx*). Male lemmings in these genera are heterogametic, but some females possess an X variant (denoted X*) causing both X*X and X*Y to be female (Fredga, Gropp, Winking and Frank 1976,

Table 6.2: Life history differences between early-maturing and late-maturing *Meriones unguiculatus* (* denotes a significant difference — Student's two-tailed test). Data from Clark et al. (1986)

Mother matures	Early	Late	Difference
Number of litters per mother			
Born	5.2	3.0	*
Weaned	4.7	2.5	*
Mean litter size			
Born	6.0	5.3	ns
Weaned	4.8	3.2	ns
Mean number of offspring			
Born	32.7	17.4	*
Weaned	25.1	11.4	*
% males per litter			
Birth	40	50	*
Weaning	40	53	*
% early-maturing females per litter	65	42	*
Age at time of first litter (days)	118.7	161	*
Mean life-span (days)	325.0	344	*
Mean post-reproductive life-span (days)	23.3	104	*

1977; Gileva 1980; Gileva and Chebotar 1979). In some *Dicrostonyx* the Y is probably absent (some females are X*0, the males XO), though sex inheritance remains unaffected (Gileva 1980; Gileva et al. 1982). The only other mammals known to have similar systems of sex determination are members of the Neotropical rodent genus *Akodon* (Lizzaralde, Bianchi and Merani 1982; Lobato et al. 1982; Bull 1983).

Under this system, it should be possible to produce offspring of genotype YY, as occurs in the platyfish (*Xiphophorus maculatus*), which has a similar system of sex determination (Bull 1983). However, in *Myopus* the X*Y females develop an X*X* ovary through a non-disjunction in fetal oocytes. The ovary lacks a Y, and as a consequence the offspring of these X*Y females are exclusively female (Gropp et al. 1976). This has the intriguing consequence of causing both an individual and population bias in the primary sex ratio. The situation in *Dicrostonyx* is less clear. The X*O female apparently segregates X* and O in equal frequency (Gileva et al. 1982), and this should lead to the production of 25 per cent OO progeny when the female mates with an XO male. These progeny are unknown and presumed to be inviable. Because X*O females ovulate at a higher rate than XX females there is no reduction in relative fecundity. The X*X females apparently have superior fecundity to both these other types of females, presumably because they gain the benefit of both a higher ovulation rate and avoid the cost of inviability of some embryos. The higher fecundity of X*X females should drive the population secondary sex ratio to a strong female bias, and very strong biases have been reported in laboratory colonies (Bull and Bulmer 1981).

Convergent evolution to a system producing female biased sex allocation is suggested by the differences between these genera and the apparent absence of a similar system of sex determination in *Lemmus*, to which *Myopus* is more closely related than it is to *Dicrostonyx* (e.g. Anderson 1985). Lability in evolution is also indicated by the presence of species with XO females in two distantly related taxa within the Microtinae (e.g. in *Microtus oregoni*, see Ohno, Stenius and Christian 1966; Pinkel *et al*. 1982, and *Ellobius lutescens*, see Castro-Sierra and Wolf 1967; Liupanova, Vorontsov and Zakarajan 1975; Wolf, Schempp and Vogel 1979). Repeated origins of a trait in separate clades living in similar environmental circumstances is excellent evidence in support of an adaptive interpretation of the trait (Ridley 1983; Felsenstein 1985).

The evolution of female-biased sex ratios in other organisms has been well treated theoretically as a consequence of the high frequency of such biases among arthropods (Hamilton 1967; Wilson and Colwell 1981; Charnov 1982), and some impressive examples of concordance of theory and empirical data are now available (e.g. Werren 1983). However, there is some dispute over the framework in which this evolution should be interpreted. Hamilton (1967) originally developed a simple game theoretical model which suggested that if the sons of a single female compete for mating opportunities with her daughters, it may be advantageous to produce more daughters than sons. In the extreme case, a female colonising a new habitat patch need only produce a single son as additional sons will not produce additional grandchildren, but additional daughters will. This suggestion is only pertinent if the sperm supply of the male is adequate. Taylor (1981) and Grafen (1984) point out that at least two different processes are implicated in this phenomenon. First, additional sons produce fewer and fewer additional grandchildren for the mother. Second, the reproductive success of sons is enhanced because there are more females with which they can mate. Four models are available which attribute different importance to these effects. Hamilton's (1967) original model allowed for both effects. Charnov (1982) modelled a strong effect from the first benefit but eliminates the second. Maynard Smith (1985b) also retains the first effect but eliminates the second. Colwell (1981) and Wilson and Colwell (1981) suggest that the first effect *cannot* select for female-biased sex ratios and instead implicate intrademic group selection (see Wilson 1980, 1983). Recent attempts to reconcile these models weaken the generality of the group selection argument, as Charnov's model produces a stronger female bias than Maynard Smith's, in the absence of any intrademic group selection (Harvey 1985; Nunney 1985a). It is also clear that the second effect can increase the extent of female bias sharply (Harvey 1985).

It is appropriate to examine the relevance of this theory to the evolution of female-biased sex ratios in *Myopus schisticolor*. Initial attempts to examine the evolutionary stability of the X*Y condition in females indicate that invasion of an autosomal modifier or Y-linked genes which suppress X* is likely unless the population is subject to recurrent extreme inbreeding (Bengtsson 1977;

Maynard Smith and Stenseth 1978; Stenseth 1978b, 1983; Bulmer and Taylor 1980a; Carothers 1980; Bull and Bulmer 1981; Benenson 1983). The biological plausibility of the implication of extreme inbreeding is weakened by inadequate field data for these species. Bondrup-Nielsen (1985) showed that cyclical changes in density do occur in *Myopus schisticolor*, and animals are often confined to small habitat patches, from which they expand at high density. While these data do certainly not contradict the inbreeding model, they can hardly be used to distinguish *Myopus* and *Dicrostonyx* from other microtines undergoing dramatic fluctuations in numbers. Indeed, Stenseth (1978b) predicts the occurrence of female-biased sex ratios to be widespread among microtines, though perhaps controlled by a different genetic system. It is true that the presence of XO females in *Microtus oregoni* is correlated with extreme female bias in the operational sex ratio of this species (Redfield *et al.* 1978b), though the mechanism underlying both these phenomena remains obscure. Perhaps the most extraordinary observation made by Bondrup-Nielsen is that male *Myopus* survive better than females. This causes the overwintered population sex ratio to be male-biased, and means males are overrepresented in old age classes. At all other times the population is strongly female-biased, as anticipated from the sex determining mechanism.

Strategic Models of Life History Evolution

The preceding discussion should reinforce the point that it is very risky to consider life history traits in isolation. The most common escape from the problem is to attempt to predict which combinations of tactics are favoured in different habitats. The distinction between fluctuating and stable environments has played a critical role in the development of this theory (Stearns 1976, 1977). Unfortunately, the ability of theory to deal with interaction between traits and the dynamics of environmental change is uneven. For example, quantitative genetical theory appears well suited to the analysis of the interaction between traits, but the effect of environmental fluctuations remains poorly treated. Charlesworth (1984, p. 120) comments that 'The main gap in existing theory at present is in dealing with life-history evolution in temporally fluctuating environments...'

By contrast, early demographic models were preoccupied with the distinction between populations which frequently encounter conditions which favour rapid expansion, and those which are usually stable. The dichotomy underlies many general and specific models of life history evolution, including the distinction between r-, K-, A- and α-selection. The result of analyses of this kind is determined by the extent to which adult and juvenile mortality are coupled (Table 6.3). Where the prospects of juvenile survival and maturation are low, it may be advantageous to spread reproduction (Murphy 1968; Schaffer 1974b; Levin 1976; Stearns 1977; Goodman 1979, 1984; Stearns and Crandall 1981a, 1981b;

Table 6.3: The effect of variation in juvenile and adult survival on life history evolution in fluctuating and stable environments. Variable juvenile mortality reverses the direction of evolution predicted by deterministic models. Modified from Stearns (1976)

Deterministic models, or stochastic models with variable adult mortality		
Environment	Stable	Fluctuating
Development	Slow	Fast
Maturity	Late	Early
Number of broods	Iteroparity	Semelparity
Reproductive effort	Small	Large
Number of young per brood	Low	High
Life span	Long	Short
Environment	Fluctuating	Stable
Stochastic models with variable juvenile mortality		

Morris 1985). Where do microtines lie on this continuum? Although the periodic irruptions have been the major stimulus for research, examination of almost any long set of demographic data leads inevitably to the conclusion that these periods of dramatic expansion are very short-lived in contrast to the 'usual' conditions of stability or decline. The notable exception to this pattern is where colonisation habitat is very extensive in comparison to constant habitat.

Temporal variation in the prospects for survival and reproduction have led to an increasing focus on persistence as a suitable currency for assessing fitness (e.g. Cooper 1984). Bridging periods of risk also plays an increasingly important part in theory which attempts to describe the patterns of territoriality, sociality and foraging in an uncertain environment (e.g. Oster and Wilson 1978; Caraco, Martindale and Whitham 1980; Real 1980; Reichert 1981). We can recognise among this trend an interest in reduction of the variance in reproduction (Gillespie 1974, 1975; Templeton and Rothman 1974; Rubenstein 1982; but see Bulmer 1984), and an influence of paleontological studies which suggest that survival of taxa may not be strongly influenced by patterns of reproduction (Vrba 1983; Maynard Smith 1983). Persistence of genotypes is probably best measured with genealogical rather than simple genetic analyses (Cannings and Thompson 1982), but the associated theory remains weakly defined.

Spatial effects influence the outcome of all of these analyses. For example, the prospect of successful juvenile maturation will depend on the rate at which new patches become available, either as a consequence of seasonal or successional effects. Following the terminology introduced in Chapter 2 to describe habitat patterns, it can be seen that the value of F_{ann}/t and H/t will vary on a microspatial scale. I have reviewed the data of Dapson (1979) and Cockburn and Lidicker (1983) which suggests that animals are capable of responding to microspatial variation in an adaptive manner by varying the rate of maturation. Unfortunately, existing models of habitat classification do not explicitly deal with these influences. For example, Sibly and Calow's (1985) synopsis interprets life history variation according to the effect that habitat has on growth (Chapter 2), but fails to acknowledge that delayed growth may represent an adaptive response to periods of adversity or uncertainty.

These observations draw attention to the possibility that the high fecundity of the microtines is diverting our attention from an essential aspect of the selective milieu in which these animals live. This is the period when the animals are scarce. Recognition of the importance of this period allows adaptive interpretation of several parameters - the delayed growth of winter animals, changes in genotype with the seasons, adaptive shift in sex allocation, and gradients in both litter size and body size. What we do not know is which periods of adversity are most likely to be important - seasonal adversity or the period between population outbreaks. Similarly, do we interpret the capacity for colonisation in terms of seasonal increases in habitat availability, or the 'boom-and-bust' of the microtine high?

The Microtines

Three models of life history evolution which have been specifically developed to account for the patterns in reproductive performance seen in the microtines attribute great importance to fluctuation in prospects for survival and reproduction, but do so in different ways:

A shifting equilibrium point in fluctuating populations?

Schaffer and Tamarin (1973) hypothesised that the principal effect of increasing numbers will be to diminish the viability of pre-reproductive animals, citing the evidence on social suppression and survivorship to which I have already alluded. Under these circumstances, they predicted that the optimum reproductive effort will decline as density increases, as the effective fecundity of parents declines. In stable environments reproductive effort and density are driven towards an equilibrium point, provided that point is stable in time (Figure 6.4). However, where the environment fluctuates in a manner which shifts the equilibrium point back and forth between two points, or when there is a large time lag in the population's growth equation, then reproductive effort and density should conform to a clockwise trajectory around the two points of equilibrium. There are no data completely inconsistent with this hypothesis (Schaffer and Tamarin 1973; Gaines, Schaffer and Rose 1979). However, the complexity of factors determining maturation schedules and litter size mitigate against its use as a predictive model, in contrast to its obvious utility as a descriptive model. Precise quantitative extrapolations appear to be difficult (Gaines *et al*. 1979).

Fluctuating versus stable populations of microtines.

An attempt to predict the evolution of reproductive effort in fluctuating and

Figure 6.4: The relation between reproductive effort (E) and density (N). (a) Optimal reproductive expenditure E(N) declines with increasing density. The zero growth isocline is shown as N*(E). The ensuing partial oscillations push the system towards an equilibrium point (E, N*) where the lines intersect. (b) If the equilibrium point moves between two values (A and B), for example under the influence of seasonality, the point [E(t), N(t)] describes a clockwise trajectory enclosing both equilibria. (After Gaines et al. 1979; see also Schaffer and Tamarin 1973)

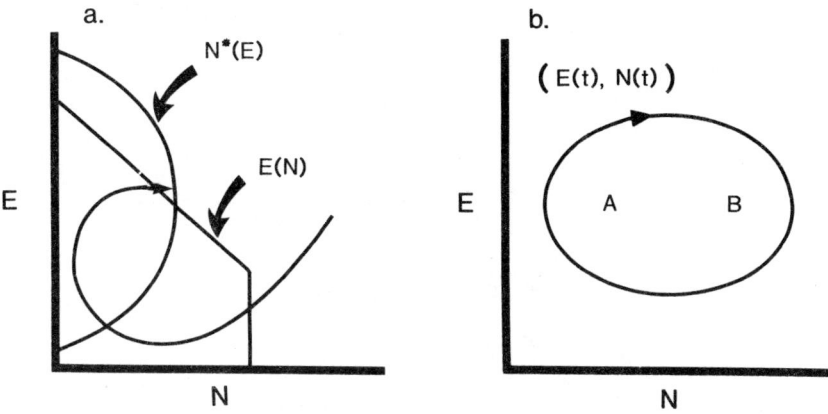

stable populations has been described in an extensive series of publications by Nils Stenseth and his co-workers Erik Framstad and Karl Ugland (Stenseth 1977, 1978a, 1980; Stenseth and Framstad 1980; Stenseth and Ugland 1985; Stenseth et al. 1985; Ugland and Stenseth 1985). In describing this hypothesis I shall not attempt to attribute the ideas to particular papers within this sequence. These papers assume a direct correlation between energetic costs and reproductive effort. Therefore their results need to be treated with caution. The authors argue that the curves relating residual reproductive value to reproductive effort will vary in form as the age of the animal increases (Figure 6.5; the authors assume that weight is correlated with age, which we know not to be true). The optimum trajectory within a range of possible fitness sets will vary according to the stability of the population. In a stable population the advantages of high fecundity are diminished, as competition assumes importance in determining the prospects of future reproduction and the prospects for survival of young. Note that, in contrast to the view expressed by Schaffer and Tamarin that optimum effort will vary with density, Stenseth is contending that the optimum reproductive effort will vary with the variance of density. In a fluctuating population, reproductive effort should peak very early, and at a higher level than in a stable population. A number of predictions can be derived from these analyses, but it is by no means clear whether the predictions are *a posteriori*, or whether the postulated curves relating RRV to reproductive effort are derived from the data they are intended to predict.

Figure 6.5: The Stenseth-Framstad life history model. The opportunity for future reproduction (residual reproductive value, RRV) will decline with age, which is assumed to be highly correlated with weight. The form of the relation between RRV and reproductive rate at each age can also be hypothesised: (a), permitting the optimum reproductive rate to be estimated for stable (unbroken heavy line) and cyclical (broken line) populations; (b), animals in cyclical populations should exhibit a higher rate early in life. (After Stenseth and Framstad 1980)

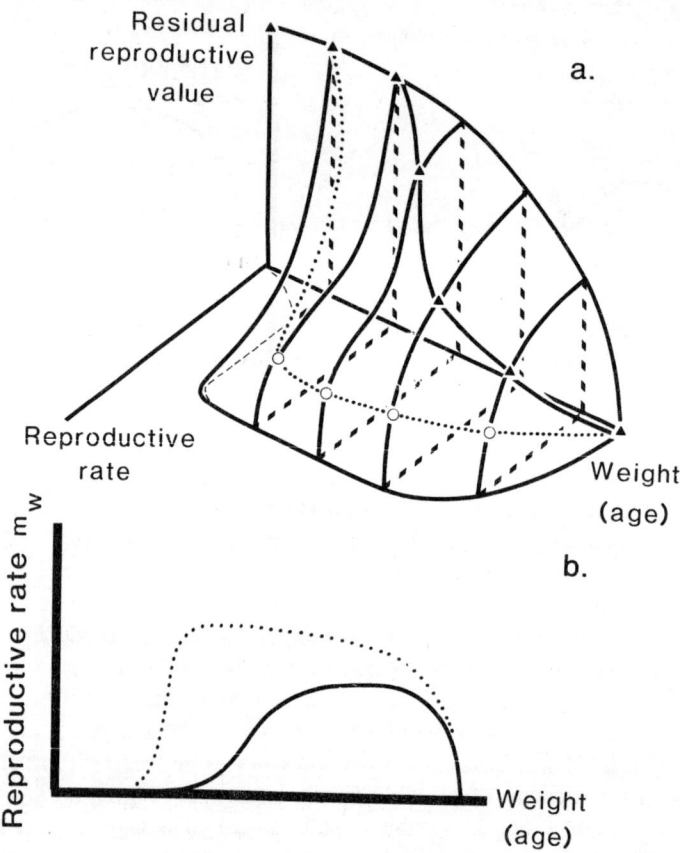

The Chitty Polymorphic Behaviour Hypothesis.

In Chapter 1, I pointed out that the Chitty Polymorphic Behaviour Hypothesis represented a model of life history evolution in response to levels of crowding. Stenseth (1978a) has pointed out most clearly the relation between the Chitty hypothesis and the model of α-selection developed by Gill (1974). Because I have already addressed the shortcomings of the original and derivative versions of the Chitty hypothesis I will not repeat the discussion here, except to say that the hypothesis can be related to the Schaffer-Tamarin model if the equilibrium point in their model is shifted back and forth by density-dependent changes in reproductive effort. By contrast, the Stenseth models only seem compatible with the Chitty hypothesis where an extrinsic factor is also implicated in the origin of cyclicity (Stenseth 1978a, 1981).

Seasonality

In summary, it is clear that life history models are capable of describing some of the changes which take place in populations which fluctuate in abundance. These analyses do not address the question of whether multiannual cycling is the dominant contributor to life history variation, or whether successional or seasonal effects are more parsimonious explanations.

Unfortunately, the life history consequences of seasonality remain very weakly explored by any theoretical approach. It is known that seasonality can affect the assumptions which underlie mathematical treatments of life history theory. Seasonality is an obvious candidate for an environmental factor which could shift the equilibrium point in the manner necessary for regular shifts in abundance in the Schaffer-Tamarin model. In addition, the most commonly used measure of fitness in a constant environment is the maximum equilibrial population size (K in a logistic model) (e.g. Charlesworth 1971; Ricklefs 1977). To rephrase this result, selection under density-dependence is expected to maximise population density. Generalisation of this expectation to other environments where density-dependence is relevant has proved problematical (Heckel and Roughgarden 1980; Hastings 1984). For example, Hastings (1984) considered two possible consequences of seasonality. When carrying capacity varies seasonally, no simple average of population size is maximised. However where density-independent mortality varies seasonally the conventional result may still apply. The outcome of more complex models is dependent on the lifespan of the organism in relation to the period habitat remains favourable for breeding (see Table 2.6).

Boyce (1979) presents a non-linear model of resource density-dependence which incorporates seasonal variation in the availability of the limiting resource. He considers a variety of dichotomies in the variation of fitness in relation to resource availability, where the important parameters are r, the maximum population growth rate when resources are limiting, D, the mean per capita resources required to maintain a constant population size, and Z, a constant for each phenotype which determines the initial slope relating per capita growth rate and per capita abundance of a limiting resource (Table 6.4). The most important assumption of Boyce's model is that natural selection favours those phenotypes with greatest fitness *averaged* over a continuous time interval. The appropriateness of this assumption will presumably depend on the specific relation of generation time to length of the breeding season, as it precludes analysis of adaptive changes in response of the type to which I have alluded (see also Fretwell 1972). It is least appropriate when generation time is very short

Table 6.4: The four sets of conditions imposed by Boyce (1979) in an analysis of the effects of seasonality on life history evolution, and their effects on the outcome of contests between the two phenotypes (1 and 2) in constant environments. The more complex case of seasonality is discussed in the text. r is the maximum population growth when resources are limiting, and r_1 is defined to have a higher value than r_2. Two further parameters are free to vary. These are D, the mean per capita resource availability required to maintain a constant population size; and z, a phenotype-specific constant which determines the initial slope relating per capita growth rate and per capita resource availability

Dichotomy	1	2	3	4
Conditions which define the dichotomy (r_1 is always greater than r_2)	$D_1 > D_2$ $Z_1 = Z_2$	$D_1 = D_2$ $Z_1 > Z_2$	$D_1 < D_2$ $Z_1 > Z_2$	$D_1 > D_2$ $Z_1 < Z_2$
Constant Environment				
Low resource levels	r_2 wins	r_1 wins	r_1 wins	r_2 wins
High resource levels	r_1 wins	r_1 wins	r_1 wins	r_1 wins

relative to the length of the breeding season.

In a constant environment the phenotype with high r is always superior for high values of resource availability. In dichotomy 3 (where resource demands of the low r phenotype are higher than demands of the high r phenotype), the high r phenotype is superior at all resource levels sufficient to maintain the population. However in the seasonal case, the low r phenotype possesses superior fitness at high seasonality levels, even in the case of dichotomy 3. It is important to note that losses are of importance in determining fitness, as is the length of the breeding season relative to the period unsuitable for breeding. Gains in summer provide little advantage in the face of severe losses in winter. Boyce's (1979) conclusions may be summarised:

(1) Low litter sizes are favoured at low levels of seasonality. This observation echoes the central conclusions of the Ashmole hypothesis;

(2) Large body size is favoured at high levels of seasonality; and

(3) Slow growth is favoured in environments with high densities or low resource availability. In highly seasonal environments the short growth season favours those individuals with the highest growth rates, as these acquire sufficient somatic gain to withstand the period of food shortage and competition.

(4) The results are very sensitive to variation in Z, leading Boyce to conclude natural selection acts principally on Z under high seasonality, because the rate of decrease during periods of resource shortage is largely determined by the magnitude of Z.

(5) If the period of resource shortage is of predominant importance, then greatest fitness may be realised by the genotype that maximises minimum fitness (see

also Templeton and Rothman 1974; compare with Levins 1968).

Because t will often be shorter than F_{ann}, these analyses are unsuited to precise dissection of the adaptive response of microtines to seasonality. However, they do shed light on a number of characters of importance to this discussion. First, they illustrate that the conclusions derived by Stenseth, Framstad and Ugland for cyclic versus stable populations may be derived where the fluctuations are driven solely by seasonality. Schaffer and Tamarin (1973) have always acknowledged that similar conclusions are true for their own models. The assumption that the latitudinal gradients in life history traits in Fennoscandian microtines are a consequence of multiannual dynamics may thus be premature, when a gradient of seasonality could account parsimoniously for the changes. It is obviously true that environments throughout the range of the microtines are characterised by seasonality, but the amplitude of the fluctuation will be most pronounced at high latitudes and altitudes, or in other circumstances where the change in resource availability is very abrupt e.g. Mediterranean annual grasslands.

Summary and Prospectus

The data I have reviewed provide tentative evidence that variation in sexual maturation, seasonal patterns in sex allocation (possibly modulated through different patterns of investment), and the phenomenon of seasonal generations are related in a cohesive way. These data support the notion that seasonality is a principal influence on life history evolution in microtines, and draws attention to the importance of considering both response to changing social conditions, and the prospects of bridging a period of uncertainty, in interpretation of life history pattern. Life history patterns in the microtines are much less diverse than is the case for mating systems. Unlike the mating patterns reviewed in Chapter 4, there do appear to be correlations between life history patterns and demography along the gradient of cyclicity in Fennoscandia.

Topics for Future Research

(1) A comprehensive theory exploring life history evolution in response to seasonality is required.
(2) The high heritabilities noted for litter size in some vertebrate populations are suggestive of complex genetic interactions amenable to quantitative genetical analysis. Appropriate data and analyses for microtine populations would be most welcome.
(3) Recent data suggest that patterns of sex allocation will repay greater attention. McShea and Madison (1986) report methods which will allow exploration of these questions in the field. A better comparative framework will

facilitate the incorporation of these patterns into a general theory.

(4) Convergent evolution of unusual sex determining mechanisms in several species of small rodents (particularly *Myopus* and *Dicrostonyx*) are indicative of unusual selection pressures. While it is tempting to relate these observations to a theoretical literature seeking to account for control of sex allocation in haplodiploid Hymenoptera, the intensity of inbreeding required to maintain female-biased sex ratios seems to be very high. Further interpretation will depend on field data, particularly for populations at low densities.

Chapter 7
Dispersal

Dispersal is any movement by an animal which leads to its permanent departure from the core area of its home range. Among behavioural characteristics which influence population dynamics, the study of dispersal is sufficiently important to warrant a chapter of its own, and has been implicated in almost every recent review of microtine demography. In this discussion, I first review the evolutionary and population consequences of immigration and emigration, and then examine the social and adaptive pressures which may prompt an animal to leave the area in which it was born, or where it has bred. In doing this I will draw on previous classifications of dispersal phenomena, and the terminology I will use is summarised in Table 7.1.

The methodological difficulties in measuring dispersal by small mammals within these classifications have proved formidable. I should say at once that I do not believe that these difficulties have been overcome in *any* study of microtines, and therefore most of the conclusions which have been drawn about dispersal in this group must be treated with extreme caution. Therefore I intend to draw heavily on studies of other taxa which illustrate the utility of evolutionary approaches to the analysis of dispersal, in the hope of illustrating profitable avenues for research on microtines, instead of repeating other reviews which cover the data which are available (Stenseth 1983; Lidicker 1985b; Krebs 1986). It is necessary to preface this discussion with a brief review of the failures of the methods most commonly applied to small rodents.

Dispersal may enhance the fitness of an animal in two ways: by changing either the habitat in which it lives, or the animals with which it associates. Adaptive interpretation of dispersal therefore requires an understanding of hetero-

Table 7.1: Classification of terms

Movement	between areas dispersal	between social groups[1] transfer
Departure	before breeding natal dispersal	after breeding[1] breeding dispersal
Population status	$r > 0$ presaturation adaptive	$r \leq 0$ saturation[2] non-adaptive[3]

Source: [1] Greenwood and Harvey (1982); [2] Stenseth (1983); [3] Lidicker (1975; 1985a)

geneity in the physical, genetical and social environments to which animals are exposed.

Methodological Problems

Because indices which are used to identify dispersers from residents are often inadequate (see Dobson 1981; Holekamp 1984a; Tamarin 1984; Cockburn, Scott and Scotts 1985, for a discussion of problems), experimental intervention has often been used to identify movements. The usual technique is to measure movement onto plots from which all animals have been removed. Immigrants are then assumed to represent a random sample of dispersers. This technique relies on the artificial creation of habitat heterogeneity, or a 'dispersal sink'. All recent studies which compared the results of a removal plot study with actual intrapopulation movements suggest that results obtained need to be treated with some caution (Krohne, Dubbs and Baccus 1984; Tamarin, Reich and Moyer 1984; Bondrup-Nielsen and Karlsson 1985; Krohne and Miner 1985; Boutin et al. 1985). The most damaging results were provided by Dobson (1981), who showed that movement by the ground squirrel *Spermophilus beecheyi* onto a removal plot was female-biased, while immigration onto unperturbed plots was male-biased. Information concerning which animals move into vacant habitat, or dispersal sinks, may be the chief benefit of this approach.

Second, in assessing sex differences or other trends in dispersal it is necessary to use the source population as a reference point. For example, the sex ratio of microtine populations is often strongly female-biased, so an excess of females among dispersers should not be taken to indicate female-biased dispersal (Pokki 1981). Unfortunately, dispersal may itself contribute to the biases in the sex ratio of the source population, so unless dispersal and mortality can be distinguished, circularity may ensue. The problem of choice of reference point is most severe when attempting to relate dispersal behaviour to some aspect of population growth or density (see, for example, Stenseth 1983), particularly when these parameters are themselves habitat dependent (Cockburn and Lidicker

1983). If we are interested, for example, in changes in natal dispersal through time, it is necessary to assess what proportion of young individuals disperse, and what proportion are philopatric. It is hardly of interest to say that natal dispersal is unimportant in declining populations if the number of young born during this period declines dramatically, as is almost always true. The best reference point is probably the proportion of weanlings which disperse, but appropriate data are unavailable for most small mammal populations.

A more intractable difficulty is illustrated in a study of dispersal by great tits. Greenwood and Harvey (1982) point out that a measure of distance moved has little biological meaning unless it is related to the number of home ranges traversed, and that the size of home ranges will depend on both population density, the extent of territoriality, characteristics of the animal including body size, diet and phylogeny, and the geometry of the habitat in which the animal lives (Chapters 2, 3, and 4). The density dependence of home range size will obviously confound the distinction between saturation and presaturation dispersal where home range fluctuates strongly with density, or where dispersal movements are small relative to home range diameter. For example, Cockburn (1985) draws attention to the effects of density dependence on the outcome of dispersal to find a genetically appropriate mate. The necessary dispersal distance will decline as the population expands and adjacent demes fuse, and the probability of these movements being detected by removal plots will also diminish.

Despite these difficulties, the study of dispersal occupies a central role in the analysis of the demography of small mammals, because its consequences for population and evolutionary dynamics are profound. Demographic consequences have been alluded to on several occasions and will be discussed in depth in Chapter 8.

The Evolutionary Significance of Dispersal

Effective dispersal can have a profound influence on the genetic structure of populations, as it determines the rate at which genes may flow through a population, and thus sets a limit to how much genetic divergence can occur between subpopulations in response to selection or random genetic drift (Endler 1977; Barton 1986). Fusion of populations through migration or expansion of the area colonised also reduces the frequency of homozygous genotypes (Wahlund 1928), which may influence the genetic consequences of different mating systems. Migration can depress local adaptation, but may also have an important 'creative' role in spreading adaptative gene combinations in species where dispersal between demes is restricted to periods of population irruptions (Wright 1978, 1982; Slatkin 1985). It is noteworthy that the spatial structure of small mammal populations is again of fundamental significance in interpreting the genetic consequences of dispersal.

If dispersal tendency is influenced by genotype, persistent emigration which is not balanced by immigration may profoundly affect the genotypic composition of populations. This observation, coupled with the demographic significance of dispersal, has led to optimism among some authors that the Chitty Polymorphic Behaviour Hypothesis could be operating through variability in the propensity to disperse (Chapter 1).

The Adaptive Basis for Dispersal

Despite the emphasis on the consequences of dispersal, it has proved much less tractable to analyse why individual animals should depart from their home range, particularly as the imbalance of emigration and immigration suggests that dispersal often leads to mortality. Dispersers may be more susceptible to predation, may carry more parasites, and may often be in poor condition (reviewed by Lidicker 1975). Whether poor condition leads to weakened animals being expelled, or is a consequence of the risk of movement, will normally be difficult to disentangle.

In this review I examine theory and empirical information pertaining to three forms of dispersal and transfer frequently reported from small mammals: abandonment of the nest by the mother (breeding dispersal), male-biased natal dispersal, and transfer between winter nesting aggregations. The widespread but non-ubiquitous occurrence of these patterns makes them particularly amenable to adaptive interpretation. I conclude that they may all be interpreted in terms of the local interactions between family members, and show that this possibility is not considered in current theory and experimental design used in research on small mammals.

Breeding Dispersal by Females

One of the most perplexing observations to emerge from descriptive studies of microtine dispersal is a high frequency of breeding dispersal by females during the breeding season. Females from several species abandon their breeding home range, or perhaps the natal nest, to their offspring (Table 7.2). This behaviour has never been investigated experimentally or properly quantified, but it appears to be most common when populations are increasing rapidly (Stenseth 1978a, 1983). Here I attempt to show that such behaviour is related to variation in the durational stability of the habitats in which these species live.

There is abundant evidence that all environments are perceived by small mammals as patchy (see Chapter 2). All field workers are familiar with consistent trap success at certain sites in contrast to those only a short distance away, and the preferential use of 'hotspots' by researchers concerned with maximising sample size is a major factor confounding the analysis of both dispersal and habitat preference. It is important to recognise that an environment which is patchy in space but stable in time may provide much the same opportunities for

Table 7.2: Species for which there is evidence that mothers abandon the brood nest (breeding dispersal) as a form of parental investment in their offspring

Microtinae[1, 2]
Lemmus lemmus
Microtus pennsylvanicus
M. ochrogaster
M. agrestis
M. oeconomus
Clethrionomys glareolus
C. rutilis
Arvicola terrestris
Dicrostonyx groenlandicus
Hydromyinae[3]
Pseudomys shortridgei
P. novaehollandiae
Sciuridae[4]
Spermophilus columbianus

Source: [1] Jannett (1980); [2] Stenseth (1978a); [3] Cockburn, Braithwaite and Lee (1981); [4] Harris and Murie (1984)

dispersal as one which is spatially homogeneous and stable in time. There are theoretical reasons for believing that animals in temporally stable environments should distribute themselves so that their fitnesses in different habitat types will be approximately equal (Fretwell and Lucas 1970; Fretwell 1972; Rausher 1984; but see Parker and Sutherland 1986). A number of observations support this concept of an evolutionarily stable 'ideal free' distribution (e.g. Whitham 1980; Power 1984), despite the formidable difficulties in designing appropriate tests. The importance of this theory for dispersal is that an animal need not increase its fitness by successful dispersal from a low quality patch to a high quality patch.

When temporal changes in habitat quality are superimposed on a pattern of spatial heterogeneity, the advantages of dispersal change markedly. Gadgil (1971) suggested that several aspects of habitat heterogeneity would promote the evolution of dispersal tendency. These include low durational stability of habitat patches relative to the species' generation time, close proximity of highly suitable patches, and catastrophic mortality creating vacant habitat patches. Variance in the ability of individuals to control local resources will exaggerate this effect, producing very low thresholds for the evolution of dispersal tendency (Lomnicki 1978, 1980, 1982; Uchmanski 1983). Extrapolations of this model to microtine life histories are discussed at some length by Stenseth (1983).

Movements of microtines can be usefully described in terms of the descriptors of habitat heterogeneity introduced in Chapter 2. Movements into colonisation:reception habitat, into most invasion habitat, and failure to traverse hostile areas between two habitable patches, lead to the dispersal sink described by Lidicker (1975). Although these movements may have considerable economic significance (e.g. Hansson 1986), and demographic significance (Lidicker 1975), their adaptive importance is negligible (Anderson 1970).

An animal which lives in a patchy environment which disperses from a constant habitat to colonisation or invasion habitat, will only benefit if breeding is possible in this new patch (colonisation:production habitat). Such movements will only be of evolutionary significance if reproduction is sufficiently prolific to ensure that at least some offspring recolonise patches constantly suitable for occupancy (Chapter 2, Figure 2.4).

Animals which live in a seasonally ephemeral habitat may approach the problem of recolonisation in two ways. If reproduction is extremely rapid, and a large proportion of a patch of ephemeral habitat is colonised by the descendants of a single female, the prospects of successfully infiltrating at least a few descendants back into the favourable habitat are good if the area of ephemeral habitat is sufficiently large. Saturation of a large patch by the descendants of a single female is only likely if the female colonises the patch early in the period during which it is available. Under these circumstances successful saturation may ensue if the female invests little time in each litter, abandoning them soon after weaning and moving to a new site to breed again. It is worth mentioning that females should have little difficulty procuring mates given the high flux of immigrant males at this time (see below). Although the mother provides little post-weaning care, she may invest through the provision of a breeding site for her daughters, leading to eventual saturation of the patch. As the prospects of successful movement by adult females diminishes, and the need to relocate into a constant habitat increases, the optimum strategy will be to enforce dispersal of all young, to enhance the probability that at least some will successfully relocate. The prospect of at least one young surviving is greatly increased if this dispersal can be spread through time (e.g. Giesel 1976). Reproduction should be extended, particularly where there is no prospect of surviving the season of greatest stringency (see Chapter 5).

Behaviour of this sort is not unique to microtines. Although populations are typically extremely stable, dispersal by female *Antechinus* appears to be most common in populations recovering from drought, or experimental manipulation (Table 7.3). Harris and Murie (1984) demonstrated that female Columbian ground squirrels (*Spermophilus columbianus*) often abandoned their nest site to a daughter, but retained their nest if yearling daughters were absent. This appears to be a form of parental investment, as mothers are dominant to their daughters and should be able to exclude them, and there is little evidence of benefit and some evidence for cost associated with abandonment. Although data were not available for populations which differed in density, Harris and Murie speculated that the low density at their study site contributed to the prevalence of nest abandonment.

Cockburn, Braithwaite and Lee (1981) showed that local survival of adult and juvenile *Pseudomys shortridgei* changed dramatically with time since fire

Table 7.3: Number of *Antechinus stuartii* females breeding on 7.5 ha study grids, and the relative numbers of immigrants and daughters of those females which bred the following year. Breeding densities had fluctuated because of experimental manipulation or as a consequence of extreme drought. Breeding densities are usually very stable

	Breeding number	Number of immigrants	Number born on grid	Prop. of immigrants
Sherbrooke Forest				
1982	5	10	6	0.63
1983	16	0	22	0
Monga Forest				
1984	11	5	13	0.27
1985	18	1	34	0.03

and suggested that this was a reflection of the relative frequency of breeding and natal dispersal. This species depends on highly productive heathland which has burnt recently and disappears from mature habitat as productivity and seed production decline (Cockburn 1978). In deteriorating habitat, adults are philopatric and long-lived, often breeding in each of three consecutive years. Young are forced to disperse from their natal home range. In recently colonised habitat, adults rarely survive beyond their first breeding season, but their offspring are philopatric and breed successfully in or near their natal home range. Cockburn *et al.* (1981) suggest that in recently burnt habitat adults should maximise juvenile survival, and may have a reasonable probability of successful dispersal to an uncolonised site nearby. In mature habitats, adults should enforce dispersal of young for as long as possible to minimise the probability that they will leave no young at all. This would occur if their home range was burnt, or no new habitat became available nearby, or if the productivity of seeds continues to decline, supporting a smaller and smaller resident population. Although the habitat in which *P. shortridgei* occurs may initially appear to have little in common with that occupied by species which fluctuate in a cyclic manner, evidence abounds that both microtines and snowshoe hares live in a spatially heterogeneous environment where the durational stability of patches is highly variable (Chapter 2).

These observations from unrelated rodents suggest that temporal fluctuations in habitat suitability may influence the relative frequency of natal and breeding dispersal. In benign or improving conditions, breeding dispersal, particularly by females, may be more probable than when the population is crowded or conditions are deteriorating.

Sex Differences

In common with many simple but illustrative models, the original mathematics used to derive the unsurprising conclusion that dispersal should be favoured in fluctuating environments were based on haploid (asexual) populations (see Stenseth 1983 for a review). The plausibility of the results may be one reason why the effects of sex differences have only rarely been included in formal theory (e.g. Motro 1982b), which therefore makes no predictions about which sex should be the predominant disperser. However, in the majority of bird and mammal populations one sex tends to move earlier and further than the other (Greenwood 1980, 1983; Table 7.4). Further, there is preliminary evidence that sex-biased dispersal confounds interpretation of genetic structure of populations, by creating local heterozygote excess when migration occurs before mating (Prout 1981; Kawata 1985b).

There are several suggestions why one sex should disperse more frequently than the other. Most of these arguments centre around the desirability of choosing a genetically optimum mate, and differences between the sexes in the nature of selection of breeding sites. High rates of inbreeding are a direct consequence of philopatry of both sexes, and any philopatry will impose a spatial structure on a population unrelated to the extrinsic characters of its habitat. Under these conditions relatedness will be proportional to geographic separation. Dispersal of one sex should help prevent close inbreeding, which tends to reduce fitness (Falconer 1981), presumably because deleterious recessive alleles are more likely to be expressed, and beneficial interactions between different alleles at the same locus (heterosis) will be lost. The extent of these effects in natural populations are controversial, particularly as any deleterious alleles should be lost very quickly, hindering the evolution of complicated inbreeding avoidance systems (Shields 1983).

Although inbreeding avoidance provides a possible theoretical explanation for sex-biased dispersal, it does not really predict which of the sexes should disperse, though the data in Table 7.4 suggest that this is reasonably predictable in both birds (females) and mammals (males). Greenwood (1980, 1983) pointed to a correlation between mating systems and the direction of dispersal. Most birds exhibit obligate monogamy, and the male must establish and maintain a territory to which he attracts a mate. Dispersal to a new breeding site may be risky, as the disperser risks trading a territory which is known to support reproduction for one which does not (or no territory at all). Females, by contrast,

Table 7.4: The number of species and families of mammals and birds with male-biased, female-biased and no sex difference in natal dispersal. Modified from Greenwood (1980)

| | Predominant dispersing sex | | | | | |
| | Mammals | | | Birds | | |
	Male	Female	Both	Male	Female	Both
Species	45	5	15	3	21	6
Families	23	4	7	1	11	5

will have a comparatively low investment in defence of resources but may benefit through dispersal by increased choice among male-defended resources. However, in the polygynous systems common in mammals the relative investment in resources tends to shift in favour of the limiting sex. In most cases this tends to be females, as the reproductive success of males will be limited principally by the number of females to which they can gain access (Chapter 3). This should lead to selection for female philopatry, as females risk most by dispersing from an area which supports reproduction. However, males may benefit through increased access to females by dispersing, particularly as young males will be unlikely to compete successfully with older males.

Male-biased natal dispersal.
Although microtine dispersal has been reviewed on a number of occasions (Lidicker 1975; Gaines and McClenaghan 1980; Tamarin 1980; Stenseth 1983), and there is general agreement on the demographic consequences of dispersal, there is no evidence of any real consensus over its causes or degree of concordance with theory. Indeed, there remain some authors who express scepticism as to whether dispersal is interpretable within an adaptive framework at all (Anderson 1980; Shields 1983). This is no doubt partly due to the methodological difficulties described above and the near impossibility of observing the proximate behaviour which underlies dispersal movements in these secretive animals. Despite these shortcomings, there is good and growing evidence that natal dispersal occurs more frequently than breeding dispersal, and that males are overrepresented among natal dispersers in many species of microtines (Table 7.5). This pattern is repeated throughout the mammals, and there are no good reasons to believe that its occurrence in microtines should be interpreted within the idiosyncratic selective milieu of cyclic or fluctuating populations. Among grouse, females apparently disperse before, and further than, males (e.g. Keppie 1979; Jamieson and Zwickel 1983; Schroeder 1986), consistent with what is known for other avian species (Table 7.4).

In order to examine the hypothesis that social interactions in a source population led to expulsion ofstressed voles, McDonald and Taitt (1982) measured the concentration of steroid hormones in blood plasma of *Microtus townsendii* moving onto a removal grid during a 'spring decline' in numbers. Perhaps to their surprise, they found that dispersing males, most of which were young and possibly reproductively inhibited, showed no signs of stress. The same was not true for females, which towards the end of the decline frequently had plasma total corticosterone concentrations in excess of their maximum corticosteroid binding capacity (Table 7.6). The significance of this parameter as a measure of stress is elaborated by Lee and McDonald (1985). These results suggest that the conditions underlying dispersal of juvenile males are unrelated to social pressure, and differ from the causes of female dispersal. Rasa and van den Höövel (1984) have also suggested that in captive colonies of *Microtus arvalis*, small (subordinate?) females are more susceptible to hypoglycaemic shock than are males.

Table 7.5: Sex bias in natal dispersal among microtines according to the mating system. Note that there is virtually no unambiguous documentation of juvenile dispersal for microtines. Data for communal nesting and resource defence polygynous systems where females are territorial are particularly unsatisfactory

Resource-defence polyandry		
M. pinetorum	Male bias	FitzGerald and Madison (1983)
Monogamy		
M. ochrogaster	No difference	Gaines and Johnson (1984)
Resource-defence polygyny (females non-territorial)		
M. agrestis (island population)	No difference	Pokki (1981)
M. californicus	Male bias	Riggs (1979)
M. xanthognathus	Male bias	Wolff and Lidicker (1980)
Male-dominance polygyny		
C. glareolus	Male bias	Mazurkiewicz and Rajska (1975)
C. rufocanus	Male bias	Kawata (1985b); Saitoh (1983)
M. pennsylvanicus	Male bias	Dueser, Wilson and Rose (1981)
	Male bias or equal	Tamarin et al. (1984)
	Male bias or equal	Keith and Tamarin (1981)
Mating system unknown		
M. breweri	Equal	Tamarin (1977a)
M. townsendii	Male	Krebs et al. (1976)
Synaptomys cooperi	Male bias or equal	Gaines, Baker and Vivas (1979)

Table 7.6: Total corticosterone and maximum corticosterone binding capacity (MCBC) (μmol/L) in the plasma of male and female *Microtus townsendii* moving on to a removal grid during a spring decline in numbers. Data from McDonald and Taitt (1981)

	Male		Female	
	Corticosterone	MCBC	Corticosterone	MCBC
16 April	0.72	1.12	5.9	7.2
29 April	0.84	1.27	7.6	6.2

The ability to distinguish whether local competition for resources, competition for mates or inbreeding avoidance are principle contributors to sex-biased dispersal in higher vertebrates has been hampered by the absence of suitable study organisms. Dobson (1982) contends that all three factors are probably involved, but attributes particular importance to intrasexual competition for mates. Holekamp (1984a, 1984b) uses similar data to implicate inbreeding avoidance. Wolff and Lundy (1985) have successfully demonstrated that sex-biased dispersal will act to reduce the probability of inbreeding in *Peromyscus leucopus*, but were unable to prove causation. Liberg and von Schantz (1985) suggest that kleptogamy will be a major form of intrasexual interference leading to male bias in dispersal. Moore and Ali (1984) suggest that intrasexual competition is a parsimonious explanation in all published instances where inbreeding avoidance has been implicated as the cause of sex-biased dispersal (see Packer 1985 and Dobson and Jones 1985 for a refutation). However, a recent study by Cockburn, Scott and Scotts (1985) using the unique life history of the marsupial genus *Antechinus* suggests that this scepticism is unwarranted. Male generations of *Antechinus* are discrete, and punctuated by an abrupt post-mating mortality of all males. Mothers always enforce dispersal of all of their sons immediately after weaning, but tolerate their daughters, and nest communally with them for several months. There is no experimental evidence which distinguishes whether mothers recruit unrelated males to live in the female kin groups, or sons use these groups as a cue for post-dispersal settlement, but circumstantial evidence suggests that both effects are important. Competition for mates among males should not promote complete dispersal by males, particularly as there is no class of experienced males. There is no risk of kleptogamy of the type proposed by Liberg and von Schantz (1985) in *Antechinus*. Any maternal advantage is unlikely to accrue from reduced competition for resources, as group size does not change significantly as a result of natal dispersal (Figure 7.1). Although I have no data confirming the deleterious effects of inbreeding (it has never occurred on

Figure 7.1: Mean number of male and female Antechinus *sharing communal nests at Sherbrooke Forest, Victoria. Upper shaded area, adults; Lower shaded area, juveniles living in the nest in which they were born; Unhatched areas, juvenile not born in the nest. Data from Scotts (1983) and Cockburn, Scott and Scotts (1985)*

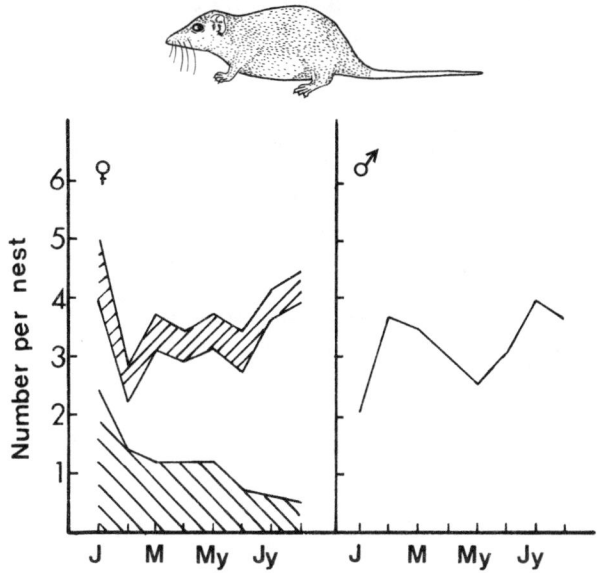

my study sites), incest avoidance seems to be a parsimonious and sufficient cause for the evolution of male-biased dispersal in *Antechinus* (Cockburn, Scott and Scotts 1985). Males do not move to nearby nests, but cross several home ranges before they are recruited, suggesting inbreeding with relatives other than siblings is also avoided.

Dispersal to avoid inbreeding seems to suggest that the costs proposed by Bengtsson (1978) are insufficient to counter the benefits obtained. This conclusion is consistent with growing observations of deliberate incest avoidance, often mediated by movement within the population (e.g. in *Meriones unguiculatus*, Ågren 1983, 1985; *Cynomys ludovicianus*, Hoogland 1982; *Melanerpes formicivorus*, Koenig, Mumme and Pitelka 1984; reviewed by Harvey and Ralls 1986). Bateson (1983) and Shields (1983) have argued that there are several costs associated with choosing a mate on the basis of genetic dissimilarity. In particular, local adaptation will break down when favourable gene complexes are scrambled. In addition, there may be costs of outcrossing which counterbalance the costs of inbreeding. These are summarised in Table 7.7 (see also Bateson 1983; Shields 1983). In particular, Bengtsson (1978) has suggested that low population density and patchy distribution will lead to very high costs for dispersal to achieve optimum mate choice. Such conditions are characteristic of microtine rodents during much of the population cycle. These costs and benefits have lead Bateson to propose the concept of optimum outbreeding, for which some supportive evidence is available from both birds (Bateson 1982, 1983; Bateson, Lowick and Scott 1980; McGregor and Krebs

Table 7.7: The hypothetical disadvantages of inbreeding and outbreeding. From Bateson (1983). In any one species only some of these costs would be likely to operate. Note particularly cost of outbreeding (item 5), and see also Bengtsson (1978)

Costs of inbreeding
(1) Deleterious genes are more likely to be fully expressed
(2) Beneficial interactions between different alleles at the same genetic locus (heterozygous advantage or overdominance) are likely to be lost
(3) Offspring insufficiently variable for one or another to cope with a varying environment
(4) Offspring more like each other and so compete more intensely

Costs of outbreeding
(1) Genes required for adaptation to particular environment lost or suppressed
(2) Co-adapted gene complexes broken up by recombination
(3) In polygynous species advantage of having extra-closely related offspring lost and parental genes less well represented in next generation
(4) Infection from pathogens carried by mate more likely
(5) Dispersal may be costly and dangerous
(6) Acquired skills useful in one environment and not appropriate in another
(7) Mismatch of habits acquired by mates in different environments disrupts parenting

1982; Grant 1984; see however Millington and Price 1985) and plants (Price and Waser 1979; Waser and Price 1983). In order to probe this possibility further, I have recently examined dispersal near a hybrid zone for teat phenotypes of *Antechinus stuartii*. The study site is located where plateau adapted animals with ten teats and coastal adapted animals with eight teats interact and hybridise. Teat number in *Antechinus* is almost perfectly correlated with litter size, and varies along climatic gradients, but in most sites intergradation is very gradual (Cockburn, Lee and Martin 1983). Animals with nine teats have low but variable weaning success in comparison to eight and ten teat animals (Figure 7.2), and consequently have the characteristics of low fitness hybrids (Endler 1977).

Figure 7.2: Number of weaned granddaughters descending from female Antechinus stuartii *which bred at Monga State Forest, New South Wales in 1983. Females with eight teats are coast adapted, and those with ten teats are plateau adapted (see also Cockburn, Martin and Lee 1983). Females with nine teats only occur in a narrow hybrid zone where the two phenotypes are parapatric. Mothers normally saturate their teats with offspring*

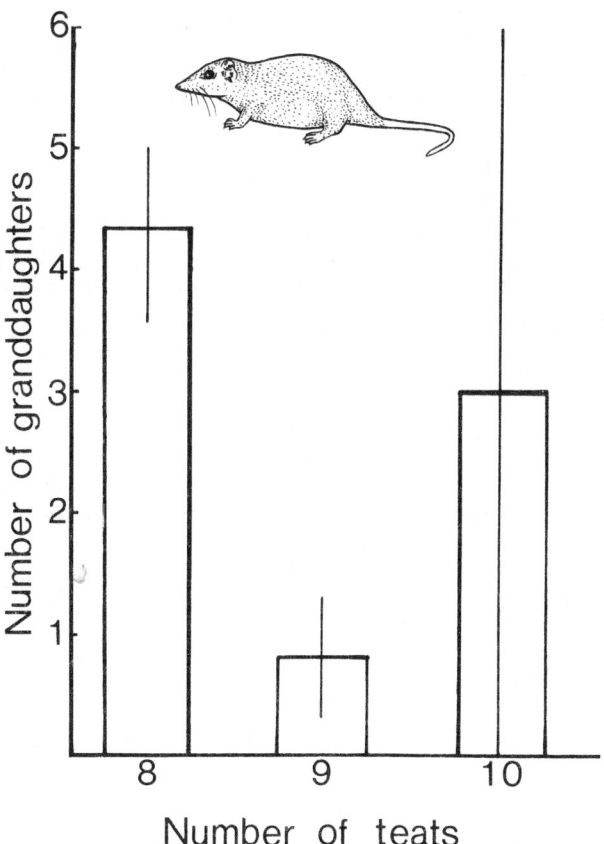

Available data suggest that hybridisation is rare except when population density is very low, directly counter to Bengtsson's (1978) prediction. Effective dispersal by males across the hybrid zone at other times is negligible, although these are the movements most likely to be revealed by the configuration of the study plot, and in spite of the large movements made by *Antechinus* away from the hybrid zone (Cockburn *et al.* 1985a). These very preliminary data indicate that optimisation of mate choice according to genetic criteria may shape social evolution and dispersal behaviour in *Antechinus*, and support the contention of Bateson (1983) that optimum outcrossing may be of importance in social dynamics in a range of vertebrate taxa.

Transfer Between Nesting Constellations.

Transfer is of particular interest as movement of this type changes only the individuals with which an animal associates, rather than the habitat in which it lives. I have already mentioned that many microtines form social aggregations during the winter, presumably to reduce energetic costs (Grodzinski, Makomaska, Tertil and Weiner 1977). Widespread transfer between these nesting groups has been documented at the start of the winter in *Microtus xanthognathus* (Wolff and Lidicker 1981) and during the winter in *M. pennsylvanicus* (Webster and Brooks 1981; Madison, FitzGerald and McShea 1984). In both cases the effect of transfer is to reduce the level of relatedness within nests as family groups break up and individuals reaggregate more or less randomly (Figure 7.3), though female relatives remain more likely to nest together than male relatives.

Madison *et al.* (1984) suggest that local depletion of food resources and increased risk of detection by predators may set an upper limit to winter group size and for *M. pennsylvanicus* the optimum group size is between two and three, though their argument is confounded by the use of 'what is must be optimal' logic, and is thus circular (Gould and Lewontin 1979). There now appears to be consensus that optimum group sizes will rarely be attained in nature, because of conflicts of interest, and variance in costs and benefits associated with joining a group, or expelling an animal from a group (Pulliam and Caraco 1984). Certainly in the case of *M. xanthognathus* individuals in winter groups cooperate to cache food, which is likely to compensate for any costs of local food depletion (Wolff and Lidicker 1981).

Such benefits shed little light on the reason for transfer between nesting constellations. Transfer appears to offer benefits to 'cheaters' which succeed in placing offspring in a nearby nesting constellation, but do not reciprocate, expelling any intruders from their own social group. Wolff and Lidicker (1981) hypothesise two chief benefits:
(1) spreading the risk of predation (see also Rubenstein 1982); and
(2) avoiding inbreeding.
They argue that discovery of a nest by a predator (usually mustelids in the case of *M. xanthognathus*) could lead either to total destruction of a nest or reduction

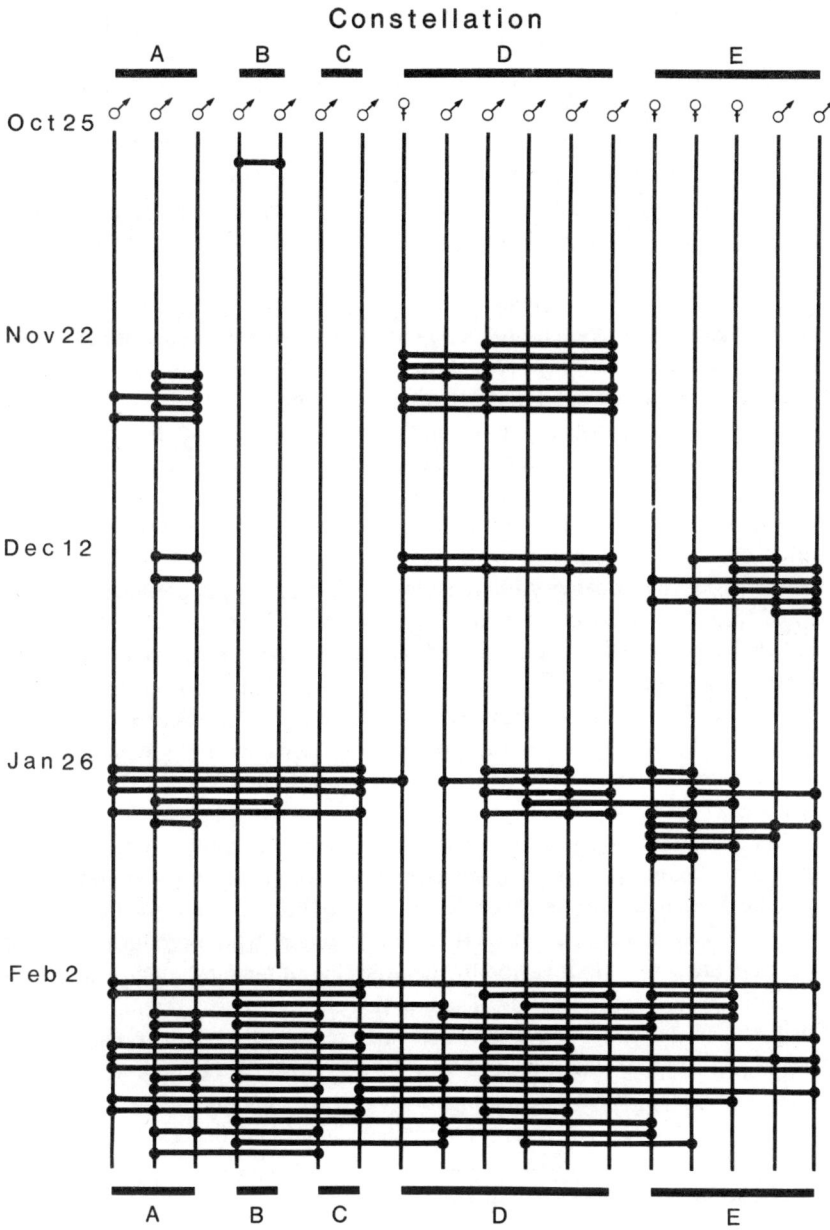

Figure 7.3: Association by Microtus pennsylvanicus *living in an enclosure into five nesting constellations. Nest use is shown by the circles on the vertical lines. Circles joined by horizontal lines indicate nest sharing. Animals nested exclusively within one constellation through 12 December, but after that time there was substantial interchange. Modified from Madison, FitzGerald and McShea (1984), to exclude animals which did not survive the entire study period*

in group size below that which would facilitate energy saving through huddling. By dispersal of some family members, there may be some individuals which survive the winter. Although we lack conclusive evidence for microtines, winter transfer and aggregation have been documented in *Antechinus stuartii* (Scotts 1983; Cockburn, in preparation), where these explanations are unlikely. In these dasyurid marsupials there is an abrupt long-distance dispersal of all males immediately after weaning, eliminating any possibility of nuclear family incest (Cockburn, Scott and Scotts 1985; see above). However there is substantial subsequent transfer between nests which gradually reduces the number of related females associated with a nest, without reducing group size (Figure 7.1), suggesting that inbreeding avoidance is unlikely to be relevant in this case. Nests are generally located in small hollows in dead trees or fallen logs which are unlikely to be accessible to predators (mainly snakes and feral foxes and cats). Long-term studies of occupancy of these nests have failed to reveal a single instance of successful nest predation. Cessation of occupancy of a nest is chiefly caused by abandonment of the site when female groups break up at the time of mating, and mothers find a new nest in which they can rear their young (Cockburn, in preparation). The ability of mothers to locate new natal nests indicates that nests are not a limiting resource during winter.

In the following section I develop theory which suggests that transfer may function to achieve a dispersion of daughters which reduces local competition among female relatives, but increases the risk of competition with neighbouring matrilines. Evidence is accumulating that as a consequence of female natal philopatry mothers compete with their daughters in many species of small mammals, leading to suggestions that where competition is prevalent, mothers should bias sex allocation towards the production of sons (Clark 1978; Taylor 1981; Silk 1984; Cockburn, Scott and Dickman 1985; see above discussion of McShea and Madison 1986). Indeed, W.D. Hamilton (pers. comm. cited in Charnov 1982) has argued that where there is strong local resource competition it can be evolutionarily stable for species to produce only a single daughter, and that additional investment in daughters may reduce the fitness of their female relatives.

One pertinent body of theory is the prediction by Hamilton and May (1978) that at least some dispersal also appears to be evolutionarily stable in homogeneous environments, where all available sites for effective dispersal are already colonised. This is because the territories of mothers which produce only non-dispersers are subject to invasion by the offspring of females which enforce dispersal of some of their offspring. The non-disperser phenotype can never replace a disperser phenotype, but the converse will occasionally be true (Hamilton and May 1978; Comins, Hamilton and May 1980). This effect occurs in both conventional population genetical and game theoretical models containing a variety of assumptions, and even where there is more than one disperser phenotype in the population (Motro 1982a, 1982b, 1983; Comins 1982). Two specific cases are of particular interest. First, Hamilton and May (1978) and Motro (1983) show that there may be parent-offspring conflict over the optimum dispersal fraction. A higher proportion of offspring should disperse if dispersal frequency is controlled by genes in the mother rather than in her offspring. Second, the result holds even when dispersal is localised to nearby sites. Comins (1982) showed that although the probability of competing with an alternative allele is reduced by the gene frequency correlations caused by localised dispersal, any effect on the optimum dispersal frequency tends to be cancelled out by the necessity to increase dispersal to seek recolonisation further afield. Unfortunately, empirical data pertaining to this theory will inevitably be confounded by the ubiquity of spatial and temporal heterogeneity in the extrinsic characteristics of animal habitat. It is important to realise that these models pertain to an intrinsic heterogeneity, caused by the presence of different 'competing' alleles or genotypes within the population.

Advantages associated with limited dispersal of daughters may provide both an additional advantage of producing daughters in species which can control sex allocation (e.g. *Antechinus*), and a 'use' for daughters in species which have only limited capacity to influence the sex ratio (most microtines). Scattering of females into nearby nests may improve the capacity of mothers or their genotypes to 'capture' adjacent habitat, by increasing the probability that they will succeed in competition with the kin clusters in nearby nests. In this context, it is relevant that mothers of both *Antechinus stuartii* and *M. xanthognathus* sometimes participate in nest constellations throughout the winter, even though there is negligible probability of their surviving to breed in the following spring (Wolff and Lidicker 1981; Cockburn, in preparation). The problem with this explanation for the existence of transfer between nesting constellations is that it seems susceptible to cheating, as its successful operation relies on a degree of reciprocity, or balance in exchange of individuals between kin clusters. This is because females within a matriline are exposed to competition for local resources with any females they accept. A matriline may 'cheat' by exporting large numbers of females, but accepting only a few. Reciprocity is certainly feasible in the close knit groups which occur, for example, among the primates and social carnivores. However it may be difficult in small mammal populations where aggregations exist principally for huddling, so that expulsion from and recruitment into the group would take place in or near the nest. However, a kin group exporting daughters may suffer thermoregulatory costs unless it recruits at least some strange females, with which it then may compete, and/or juvenile males. Male-biased recruitment may seem a way of cheating, and it is true that in *Antechinus* there is very high variance in the number of males which participate in winter nesting constellations (Cockburn, Scott and Scotts 1985). However, living with a large number of males may also have costs. Female *Antechinus* are often hurt during the mating season, apparently as a result of attempts at copulation by males. The occurrence of fur loss and wounding is patchy within and between populations and it is possible that such damage is the result of a very high local operational sex ratio, with females subject to prolonged and repeated copulations (Scott and Tan 1985).

In summary, it appears that there are both energetic reasons for winter aggregations in small mammals, and theoretical reasons for expecting a moderate level of female transfer into neighbouring nesting constellations. The costs to females of competition among males for mates may counterbalance the disadvantage of accepting unrelated females into nesting constellations. Unlike many aspects of dispersal behaviour, these ideas appear to be amenable to testing without any major methodological developments. Radiotelemetry has proved useful in revealing the composition of nesting constellations, and can also provide clues concerning the physiological condition and fate of individual animals. The fate of groups and females in populations with 'normal' and manipulated sex ratios is also testable. The data from *Antechinus* show that the spread and extinction of matrilines can be accurately quantified, provided kinship in natural populations can be described. The formal models necessary to integrate these observations are not particularly complex.

Are Dispersers Genetically Distinct?

The evidence presented in this review suggests that at least three common patterns in small mammal dispersal are amenable to adaptive interpretation. It is consequently useful to place these observations within the perspective of the literature on small mammals which has sought a genetic basis for dispersal. This literature has emerged largely as a result of the optimism that dispersal may be the means through which the 'quality' of individuals changes through the course of population cycles in microtine rodents (see Krebs 1978 for a review).

Evidence for a genetic basis for dispersal comes from two main sources: (1) quantitative genetic studies which estimate the heritability of tendency to disperse; and (2) comparison of the frequency of allozyme morphs in samples of 'non-dispersers' (the source population) and 'dispersers' (animals appearing on removal grids).

Heritability

There are no completely convincing data which suggest that dispersal tendency in vertebrate populations is heritable. The most widely cited quantitative genetic analysis is a study of great tits by Greenwood, Harvey and Perrins (1979), in which correlations between distance dispersed by parents and their offspring indicated that heritability was about 60 per cent. van Noordwijk (1984) has recently shown that their original analysis was confounded by environmental heterogeneity, and by the restrictions to sampling created by a finite study area (see also Barrowclough 1978). Correlations between dispersal tendency of microtine siblings have been reported more than once (Hilborn 1975; Beacham 1979a; see also Keppie 1980), though these studies were based on very restricted samples within families, which may have confounded their results to an unknown extent. Unfortunately, the only study which has attempted formal measurements of heritability of dispersal tendency is found in an unpublished thesis (Anderson 1975), though Krebs (1979) summarises her chief results. *Microtus townsendii* offspring of known parentage were released on to both fenced and unfenced grids, and between-sibling correlations were compared to estimate heritability (Falconer 1981). If dispersal tendency was heritable Anderson anticipated and found a higher heritability for length of residency on the unfenced grid where dispersal was possible ($H^2=0.62$) than on the fenced grid where length of residency presumably reflects the artefacts of sudden release on to an unfamiliar area ($H^2=0.37$). Although these data are consistent with some genotypic influence on dispersal, the argument is very indirect, and Krebs (1979) rightly urges caution in their interpretation. Replication of this study with other species and population conditions should be a high priority for researchers.

Allozyme Frequencies

Changes in allozyme frequencies as a result of dispersal have attracted considerable attention because such changes may underlie population fluctuations of this sort. Attempts to establish a difference between the frequency of electromorphs of enzymes in residents and dispersers have been simultaneously encouraging but equivocal. Evidence that dispersers are not a random sample of the population have been documented for *M. pennsylvanicus* (Myers and Krebs 1971b; Keith and Tamarin 1981), *M. ochrogaster* (Myers and Krebs 1971b; Pickering, Getz and Whitt 1974) and *M. townsendii* (Krebs *et al.* 1976). Overrepresentation of genotypes among dispersers was sometimes restricted to certain phases of the population cycle. For example, Myers and Krebs (1971b) found that during the population peak and decline, dispersing males had a preponderance of homozygotes for an allele of transferrin when compared to the source population, but females dispersing during a period of population increase were more often heterozygous for transferrin than the source population. In contrast to these results, no difference between allozyme frequencies between residents and dispersers were detected in studies of *M. californicus* (Riggs 1979), *M. pennsylvanicus* and *M.ochrogaster* (Verner 1979, cited by Gaines and McClenaghan 1980), and *M. breweri* (Keith and Tamarin 1981). Keith and Tamarin (1981) argued that the difference between *M. breweri,* an island dweller and *M. pennsylvanicus,* a mainland form, reflect the restricted potential for dispersal in the former species. How these patterns relate to other factors influencing allozyme frequencies is completely unknown.

Whether these results are helpful or obfuscatory remains uncertain, and even the most ardent advocates of the uses of electrophoresis urge caution in the interpretation of extremely indirect correlations of this sort. In particular, the distinction between cause and consequence is currently insoluble. Indeed, the interpretation of electrophoretic data from microtine populations is currently undergoing reevaluation from a number of perspectives (Gaines and Whittam 1980; Motro and Thomson 1982; Gyllensten 1985; Kawata 1985b). If, as I have claimed above, genealogical and genetic structure may drive dispersal machinery, instead of vice versa, some of the conventional assumptions used in population genetics may be violated (Cockburn 1985).

Behavioural studies of the propensity to disperse have not been widely attempted, but they may be confounded in a number of ways. For example, Karlsson (1984) showed that the homing ability of *Clethrionomys glareolus* depended on age and reproductive status. Young animals showed poor homing abilities. Is this a consequence of lower familiarity with local conditions, or a reflection of their greater propensity to leave their natal range? Rasmusson *et al.* (1977) reported differences in mobility in laboratory colonies of *Microtus agrestis* taken from cyclic and non-cyclic populations (see also Table 5.2), but the ability to relate these data to field conditions remains elusive.

Thus although there are data available which suggest dispersal tendency is influenced by genotype, we are some way from resolving this question satisfactorily. Furthermore, the search for genetic influences appears to be clouded by the dichotomy between saturation and presaturation dispersal. Because dispersal is prevalent during periods of population increase when local conditions are benign, there appears to be consensus that dispersal is not 'enforced', and is hence the result of some individual predisposition. The data I have documented on both sex-biased dispersal and winter transfer suggest that enforced dispersal may be both adaptive and density independent, and reflect interactions at the level of the family. The ability of dispersal to promote sexual maturation in low density populations of *Microtus ochrogaster* can also be viewed in this light. As it may *always* be advantageous to enforce dispersal of part of a litter (Hamilton and May 1978), dispersal may be subject to selection, but its heritability is unlikely to be reflected in sibling correlations, as some siblings will disperse and others will not. While differences in allozyme frequencies may reflect between-family differences, the overwhelming importance of within-family effects may confound any use of this technique.

Summary and Future Directions

The study of dispersal in microtines has suffered from a tendency to focus exclusively on correlations with aspects of cyclicity, and consequently ignores the broader constraints on the evolution of behaviour in all mammals. At least three aspects of dispersal behaviour in small mammals are amenable to 'selectionist' interpretation within a broader framework (see Charnov 1982). These are male-biased natal dispersal, post-weaning breeding dispersal by females living in benign environments, and winter transfer between nests. The occurrence of these aspects of behaviour in the groups of small mammals referred to in this review are documented in Table 7.8, together with some important aspects of the ecology of these groups. Several patterns emerge. First, there is no evidence that *any* type of dispersal is restricted to cyclic or other fluctuating

Table 7.8: Summary of types of dispersal and important ecological features of four groups of small mammals (sources are given in text)

	Microtines	Ground squirrels	*Pseudomys shortridgei*	*Antechinus stuartii*
Diet	Herbivores	Omnivores	Omnivore	Insectivore
Interannual variance in numbers	High	Low	Moderate	Very low
Periodicity	Yes	No	Decline with succession	No
Duration of life	< 1 year	> 1 year	1 year early 3 years late	1 year
Polyoestrous/ Monoestrous	Poly	Poly	Poly	Mono
Mating system	Polygyny Monogamy Promiscuity	Polygyny Monogamy	Monogamy	Promiscuity
Habitat patchiness	Extreme	Moderate	High	Very low
Breeding dispersal by females	Yes	Sometimes	Yes	Yes
Male-biased natal dispersal	Common	Yes	No	Yes
Winter transfer	Yes	Unknown	Unlikely	Yes
Proportion of litter which disperses or transfers	High	Most males	0 to 100%	High

populations. While I agree that dispersal patterns are important in interpretation of population fluctuations, I feel that the case that cyclic population fluctuations are an important selective influence on dispersal behaviour cannot be confirmed with available data. Dispersal and transfer do not appear to be influenced by diet, as has been implied by West and Dublin (1984), or degree of intergeneration overlap, as has been implied by Dobson (1982). Natal dispersal varies with environmental circumstances from nil to 100 per cent in some species (e.g. *Pseudomys shortridgei*), and is restricted to one sex in other species (e.g. *Antechinus*). The chief influences on dispersal and transfer appear, as has long been recognised, to be heterogeneity in the social and physical environment. However, genetic heterogeneity (the distribution of relatives, or of competing alleles) may be important in most mammalian populations, and may lead to an underlying pattern of dispersal which needs to be isolated before progress can be made in other areas.

Topics for Future Research

(1) These results suggest that the methods currently used in studies of small mammal populations may be unsuited to the dissection of the evolutionarily and ecologically important patterns which underly social behaviour and dispersal. First, it is unclear what aspects of dispersal are likely to be revealed by the removal plot technique. Cockburn (1985) has suggested that this method is most likely to reveal long-distance movements, and is quite unsuited to understanding the proximate social context of dispersal. In particular, the method does not distinguish between an increased frequency of dispersal, which is likely to be of considerable evolutionary significance, and increased distance dispersed, which will depend on number of home ranges traversed, genetic structure of the population, and habitat characteristics.

(2) Conspicuous species like ground squirrels and demographically tractable species like *Antechinus* have allowed progress towards resolution of several important questions, and suggest possibilities for studies of microtine behaviour. Now that a theory which relates genealogical and genetic structure is emerging, and advances in molecular biology seem likely to make genealogical structure accessible in all species, the design of research into dispersal by microtines warrants fundamental reassessment. There seems little point in continuing to interpret dispersal against a background of population fluctuation, as the connections to both the proximate behavioural, and ultimate evolutionary, causes of dispersal are often so diffuse as to be uninterpretable. While some aspects of dispersal may eventually be related to features unique to fluctuating populations, the most exciting progress seems likely to come from a synthetic approach which recognises that dispersal patterns will both influence and be influenced by elements of social organisation common to many mammalian species, and in particular by the interactions which occur within families.

(3) One aspect of microtine dispersal which does seem worthy of further investigation is the role of seasonality. Gliwicz (1986) reviews data documenting seasonal effects in several populations of small mammals, but her interpretation continues to reflect the saturation-presaturation dichotomy. Further investigation of the role of seasonal effects in promoting or discouraging dispersal will allow construction a more synthetic theory of life history evolution.

Chapter 8
Social Behaviour and Population Regulation: a Synopsis

The suspicion that population dynamics of fluctuating populations of vertebrates are a consequence of density-dependent changes in the social milieu has stimulated one of the most substantial bodies of research ever attempted by field biologists. Elton (1942, p. 205) wrote:

> Without at present putting the matter any higher than this, we can see that there are inherent properties of the population dynamics of voles that may eventually explain their cycle as a self-contained system that is not so much dependent on other animals like predators and parasites as we at first supposed. And that this system is complex and delicate, and not by any means fully understood. We have moved a long way from simple epidemiological or climatic theories into a new region of population research.

Compare this quote with the synopsis by Taitt and Krebs (1985) that I cited in Chapter 1. Can we claim to have progressed into this new region at all? There is certainly no evidence of any consensus between major contributors to the study of multiannual fluctuations, and the optimism is fading.

The Strategy of Investigation

Continued faith in this approach in the absence of any consensus among researchers suggests that there is something wrong with the way in which data are currently being integrated and interpreted. A critical examination of philosophy may warrant brief discussion, as a complement to the discussion of results already presented. Anyone who has worked on microtine biology will know that there are 'schools of researchers' who at times pay their colleagues little heed. In writing this review, I was amazed on one occasion when a paper was published which covered a topic I had just written about. The references we used in support of our similar arguments were mutually exclusive. While part of this factionalism reflects difficulty in maintaining contact with a growing literature, at least part of the problem reflects differences about what can be accepted as reliable evidence. To illustrate this problem, I offer the following summary of the sociology and philosophy of microtine research. I recognise three approaches:

The Experimental School

This group is dominated by Charles Krebs and his students. It follows the teaching of Dennis Chitty in emphasising the importance of genotypic causes of fluctuation (e.g. Gaines), and placing strong emphasis on spacing behaviour (e.g. Tamarin). The chief strength of their contribution has been an intense emphasis on *experimental* tests of hypotheses, and consequently on *explicit* statements of the hypotheses of interest (see also C.J. Krebs 1985a). I feel that the chief weakness of this approach has proved to be the difficulty of constructing definitive ecological experiments in an empirical vacuum. For example, in a decade of intensive research on spacing behaviour of *Microtus townsendii*, the pattern of territoriality has defied characterisation. Simple descriptive studies represent a much more rapid means of acquiring the relevant data to answer this *fundamental* question and facilitate the design of biologically richer experiments.

The 'It Happens, Therefore it Causes' Approach

A large part of the literature on microtine population regulation takes the form of extrapolation of individual aspects of the life history or social behaviour of a microtine species into a general theory of regulation. This approach is obviously flawed and is best illustrated by a couple of direct quotes. In selecting suitable examples, I have certainly not gone out of my way to find the most egregious cases. Instead, the proximity of the studies quoted to the top of the pile of reprints from which they were selected indicates that the studies have proved very useful to me in reviewing microtine behaviour. Nonetheless, the desire to hang results on the hook of population regulation results in conclusions which are often lame.

Winter grouping:
The thermoregulatory benefits of huddling and the risk of predation appear to be important governors of movement, group formation and dispersion. The existence of an optimum group size produces and [sic] 'Allee effect' that may contribute to population lows and multi-annual cycles. Madison *et al.* 1984.

Infanticide:
> There is still no satisfactory explanation for ... cycles. ... At low densities,... pregnancy block and infanticide would be unimportant...Given sufficient intrinsic conditions, such populations should increase rapidly. However, as density increases, so does dispersal and females would encounter increased interference by strange males ... increased incidence of pregnancy blockage, infanticide and prolonged gestation ... These density-dependent phenomena would reach a climax in peak years and contribute to a marked decline in the next season's population. Mallory and Brooks 1978, p. 146.

The Everything is Important School

Rather than propose a series of hypotheses for population regulation, other authors have sought to produce hypotheses of greater and greater complexity, which integrate all the factors which have a demonstrable demographic effect. The researchers working at the University of California at Berkeley (Frank Pitelka, Oliver Pearson and Bill Lidicker), the Finnish school dominated by Olavi Kalela and his students (e.g. Hentonnen *et al*. 1984; Stenseth 1985), and the Wisconsin group examining hare fluctuations (Keith 1974) represent convergent development of these ideas. From an early emphasis on food quality and predation, progressively more complex models have incorporated the influence of social factors such as delayed maturation and dispersal. This approach has been criticised most strongly by the experimentalists, who castigate the failure of the multifactorial hypotheses to develop quantitative predictions of the relative importance of the various factors of interest (Tamarin 1978b; Taitt and Krebs 1985). It is true some aspects of these hypotheses have defied quantitative analysis.

Is a General Theory Possible?

If a student of population regulation has persisted thus far, they are probably now white with rage and tempted to dismiss this review as a complicated evasion of the central hypothesis of interest. I therefore feel obliged to offer a few tentative suggestions of how the topics addressed in this book can be integrated into a theory of population regulation, if indeed they need to be integrated at all.

Certainly any general theory must account for the following observations:
(1) Addition of food improves reproductive performance significantly, but is less effective at preventing declines in numbers (Chapter 2).
(2) Habitat available to voles is both spatially and temporally heterogeneous. The evolutionary impact of voles which colonise areas temporarily is unclear, but such voles are the subjects which provide most of the data from which hypotheses are framed (Chapter 2).
(3) There is often synchrony between species in the timing of irruptions and declines (Chapter 1)

(4) Irruptions most closely approach true cyclicity in environments which I expect are characterised by extreme seasonality (e.g. high latitudes and Mediterranean annual grasslands), though comparative measures of seasonality have not been attempted (Chapter 1)

(5) Cyclicity in vertebrate populations occurs in some species which are quite rare, and the period of cycles is determined to some extent by allometric features. However, predation may force some species into the cycle period of the dominant member of the vertebrate community, which will usually be a herbivore (Chapters 1 and 2).

(6) Interspecific diversity in patterns of spacing behaviour (manifested through territoriality) appears to be very poorly correlated with interspecific differences in demography. By contrast, intraspecific variation in demography may be extreme, while the extent of intraspecific variation in territoriality and mating systems is unknown (Chapters 3 and 4).

(7) Variation in reproductive performance appears to show less interspecific variation than variation in spacing behaviour. There does appear to be some correlation between latitude, cyclicity and life history traits, but it remains controversial whether these correlations have a nutritional or genetic basis (Chapters 5 and 6).

(8) Microtines have a phenomenal reproductive rate, particularly given the nature of their diet (Chapters 2, 5 and 6).

(9) There is abundant evidence supporting the idea that seasonal variation in patterns of maturation is of critical importance in moulding the evolution of microtine life histories in general. However, there is little evidence that selective pressures are consistent within and between populations (Chapters 5 and 6).

(10) Although the connection between dispersal and population and evolutionary processes is clear, the proximate behavioural basis for dispersal in microtines remains unknown. Studies of other small mammal populations suggest that there is no aspect of dispersal which can be unambiguously attributed to cyclicity (Chapter 7).

Stability, Cyclicity and Chaos

There is continuous variation between populations which are 'stable', those which fluctuate regularly, and those where irruptions are unpredictable. However, claims that populations do or do not fluctuate in a predictable manner are statements with an implicit mathematical content. It is therefore useful to briefly discuss mathematical approaches to the description of population dynamics, emphasising those points least easy to extract from introductory texts in population ecology (May 1981a, 1981b; C.J. Krebs 1985a). It is useful to distinguish between three population conditions: in order of stability these are: (1) stable equilibria, where population size tends towards a constant number; (2) stable cycles, where population numbers tend to oscillate in a regular and predictable manner or variable period; and (3) chaos, where the long-term behaviour of the population is unpredictable (May 1974). The simplest descriptions of population dynamics are deterministic, and do not allow for any stochastic influences, an assumption which facilitates analysis but is of controversial biological validity. The appropriate mathematical description of dynamics depends on whether population growth is discrete and generations are non-overlapping, as is the case in many species of temperate insects; or growth is continuous and generations overlap, as is true for most birds and mammals. For non-overlapping generations, difference equations of the form

$$N_{t+1} = F(N_t) \qquad \ldots 8.1$$

where N_t is the population size at time t, are appropriate. Even very simple difference equations exhibit extremely complicated dynamics (May 1974, 1976). For example, the relation between $F(N_t)$ and N, often has the form of an inverted

Figure 8.1: Maxima in lynx pelts taken throughout Canada from 1735 to 1905 viewed as an approximately one-dimensional mapping. Successive maxima are plotted against each other. The curve or difference equation of best fit is

$X_{i+1} = 3.56 X_i^2 exp(-4.46 X_i)$, $r^2=0.83$, $df=17$

When the outlier (the data for the 1857 maximum) is excluded, r^2 rises to 0.93. After Schaffer (1985)

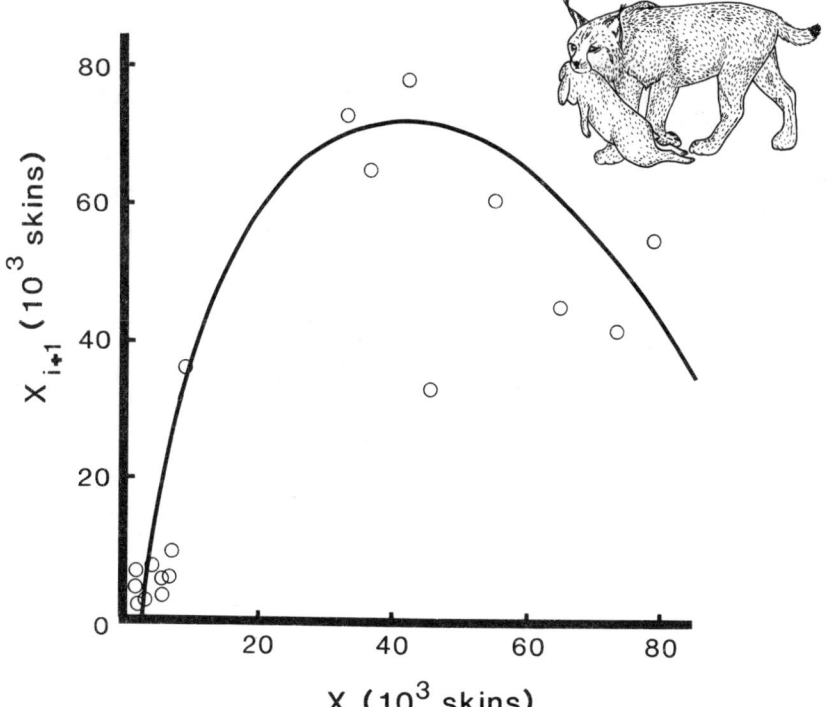

parabola (e.g. Figure 8.1; very large populations decline from one period to the next, very small populations increase). As the parabola steepens, the system exhibits a range of dynamics which vary from a stable equilibrium point through cyclic oscillations of increasing period to behaviour which is completely chaotic (May and Oster 1976). Where generations overlap, differential equations are more appropriate, but their solutions are much less tractable (May 1974), and correct definition of the appropriate equation is usually based on guesswork (Schaffer 1985). It has therefore been difficult to establish whether the chaotic dynamics of some populations depend on stochasticity, or whether the simplicity of deterministic forces is adequate for their description and modelling. The distinction between determinism and stochasticity is biologically important, as most verbal population models of cycles imply some determinism. A recent approach suggested by Bill Schaffer and Mark Kot offers some promise (Schaffer 1984, 1985; Schaffer and Kot 1985, 1986). They review recent mathematical theory which suggests that is possible to recognise order in systems with complicated population dynamics through graphical analysis. Although the precise characterisation of population trajectories necessary for analysis of differential equations will normally be impractical for field ecologists, it is possible to use approximation procedures developed for applications in statistical physics to force data series into patterns which can be treated by difference equations. While the appropriate geometry will fry the brains of the average ecologist (the current author included), a number of readable summaries are now available (I prefer Schaffer and Kot 1985). Schaffer (1985) has concluded that determinism may shape the population dynamics of lynxes, which fluctuate in abundance in synchrony with the snowshoe hares on which they prey. Although snowshoe hares fluctuate over a highly predictable time period, the amplitude of the fluctuations vary dramatically, yet can be rendered into a difference equation from which deterministic control of these chaotic dynamics can be inferred (Figure 8.1). An important corollary is that even if dynamics are deterministic, and limited prediction of maximum to maximum should be possible, long-term prediction is probably out of the question.

Demographic Consequences of Microtine Habitat, Behaviour and Reproduction

In reviewing the biology of microtines I have identified three topics of profound importance in the intepretation of their biology: habitat heterogeneity, particularly as determined by food availability; spacing behaviour, which is manifested through territoriality and dispersal, and life history patterns which appear to reflect the dominant effect of seasonality. The population consequences of all these phenomena have been subjected to theoretical analysis.

Habitat Heterogeneity

Several models have attempted to explore the relation between habitat heterogeneity and population stability, some of which deal with the properties of any fluctuating population, and some of which deal specifically with attributes (or assumed attributes) of microtine populations.

First, the relation between persistence of a population in a patch and connectivity between patches may be crucial, suggesting attention should be paid to the configuration of habitat patches. Models of these effects have been successfully applied to known parameters of the population dynamics of *Peromyscus* (Fahrig and Merriam 1985; Lefkovitch and Fahrig 1985), and spruce grouse (*Canachites canadensis* - Fritz 1979), but not to my knowledge to microtine populations. It is worth mentioning that in insect populations, a large burst of population increase, even if infrequent, may be important in maintaining population sizes large enough that extinction is improbable in bad conditions (Murphy, Launer and Ehrlich 1983). Results of this sort will in part depend on the dispersion of reservoirs of habitat after the decline.

Several models do explicitly investigate the relation between habitat heterogeneity and population stability in microtine populations. In these models damped fluctuations represent greater stability than cyclic fluctuations which in turn imply greater stability than rapid extinction. Unfortunately these models differ in their predictions concerning the relation between spatial heterogeneity and stability.

Plant/herbivore coevolution.

Rosenzweig and Abramsky (1980) suggested that predator/prey (herbivore/plant) co-evolution will only stabilise populations in a homogeneous environment. In a heterogeneous environment, spatial variability and panmixia (sustained by dispersal) are supposed to prevent voles and their food exhibiting precise coadaptation. Under these conditions some voles will be living in habitats much more productive than the habitat to which they are adapted (because panmixia will cause adaptation to an intermediate habitat type). Stability is predicted to be low in these habitats.

Responses to phenological changes.

Rosenzweig and Abramsky (1980) also argued that there should be strong selection for capacity to respond to the availability of food resources, and cite in support of their argument evidence that reproductive activation of *Microtus montanus* is facilitated by 6-methoxybenzoxazolinine, a compound found in the natural diet of this species (Negus and Berger 1977; Berger *et al*. 1981; Sanders *et al*. 1981). They postulate that there will be between-patch differences in the level of activator compound to which voles should respond. Under these circumstances migration will subvert local adaptation, and increase the probability that voles will breed when resource levels are inappropriate to sustain reproduction, particularly in patches where sensitivity should be lower than the average sensitivity of the panmictic population. The excess vole burden could harm the quality of food, and precipitate a crash. There would then be strong selection against the most prolific reproducers, as these animals would be debilitated by the costs of reproduction at an inappropriate time. This may lead to selection against sensitivity to resource tracking, and ultimately reinitiate the peak/crash cycle.

A central and testable prediction of these hypotheses is that immigration rather than emigration is crucial in the development of cyclicity, and that the breakup of local adaptation is destabilising. The elegant experiment of Tamarin et al. (1984) which showed that cycles persist in the absence of immigrants, but not when emigration was frustrated, is a simple rejection of this prediction (Figure 8.2; see also the discussion by Krebs 1986). Second, the idea that immigration occurs at sufficiently high levels to cause panmixia and preclude local adaptation is difficult to sustain. Social barriers to gene flow between demes have been documented in some small mammals (e.g. Singleton and Hay 1983), and genetic heterogeneity has been reported over very short distances in patches of habitat where there are no obvious barriers to dispersal (e.g. Bowen 1982; Krohne and Baccus 1985).

Figure 8.2: Experimental design used to examine the 'fence effect' on Microtus pennsylvanicus. *Grassland is constant habitat, but the woods do not support voles. The solid lines denote vole-proof fences. The three trapping grids (dots denote trap sites) differ in the effect they have on immigration and emigration. On Grid A emigration and immigration is frustrated. On Grid B emigration may take place into the dispersal sink of the woods, but there is unlikely to be any immigration from the woods. On Grid C both emigration and immigration are permitted. From Tamarin, Reich and Moyer (1984)*

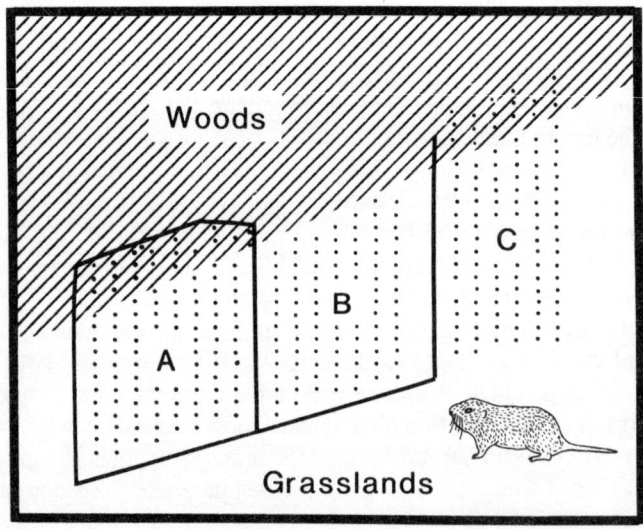

Response to predators.
In contrast to the arguments advanced by Rosenzweig and Abramsky (1980), Stenseth (1980) provided a series of models of the interactions between predators and their prey which suggest that spatial heterogeneity is likely to *stabilise* population dynamics. The case most relevant to microtine populations was an elaboration of a model by Tanner (1975) (see also May 1981b), in which two patches are defined according to predator efficiency (high, low). At low prey densities all individuals are concentrated in the environment where predation is limited, but individuals disperse into the poor patch as the carrying capacity of the good patch is exceeded. This system exhibits stability when biologically realistic parameters which generate a limit cycle between predator and prey in a homogeneous environment are used. Stenseth (1980) attempted to relate these observations to the gradient of cyclicity in Fennoscandia by suggesting that habitats at northern latitudes are more likely to be homogeneous. The biological simplicity of this single-factor model reduces its plausibility and predictive power.

Contagion and stability.
Hassell (1980) and Hassell and May (1985) have reviewed models which indicate that spatial heterogeneity acts to stabilise population dynamics under circumstances where animals exhibit a highly contagious distribution within patches. In social vertebrates, the factors which generate contagion are easy to understand (e.g. Jamieson and Zwickel 1983; West and Dublin 1984), particularly where food availability imposes an initial patchy structure on the environment. Although this theory is still poorly explored, it does provide a basis for interpretation of a number of simple observations of the dynamics of laboratory systems (Huffaker 1958; Nisbet and Gurney 1982).

Spacing Behaviour

Spacing behaviour is currently the favoured child of microtine biologists. My discussion has emphasised two forms of spacing behaviour - territoriality and dispersal, both of which are linked directly to habitat heterogeneity. Theoretical analysis is almost universal in suggesting that the role of both these aspects of behaviour will be to *stabilise* population numbers (Dispersal: see Okubo 1980; Hastings 1982, 1983b; Levin, Cohen and Hastings 1984; Vance 1984 and many others; Territoriality: Lomnicki 1978, 1980, 1982; Lomnicki and Ombach 1984).

Dispersal.
Confined populations grow more rapidly than populations from which dispersal is permitted (Lidicker 1975; Krebs 1979; Tamarin *et al.* 1984), and demography in island populations where dispersal is restricted differs greatly from nearby mainland populations (Lidicker 1973; Tamarin 1977a, 1978a; Gliwicz 1980), clearly demonstrating the demographic significance of dispersal. Dispersal in

most prevalent in rapidly expanding populations, implicating social behaviour in the limitation of population growth.

Although dispersal obviously contributes to the expansion of populations in space, it is generally believed that dispersal reduces the growth rate of populations at their core, where emigration is not balanced by immigration, and that mortality of emigrants occurs as a result of movement into a 'sink' of unfavourable habitat. If such dispersal is age- or sex-biased, the demography of the population is markedly affected, as a consequence of differences in mortality in the dispersive sex or age class. Habitat heterogeneity consequently underlies this interpretation of dispersal and its demographic consequences, although the spatial variability of environments is rarely assessed in dispersal studies.

Taitt (1985a) has suggested that although spacing behaviour will usually regulate population density because animals will be lost into 'dispersal sinks', under unusually good conditions the extrinsic factors which influence the fate of dispersers will not operate. Under these conditions many animals may move into an unoccupied area and settle at the same time. This phenomenon is called simultaneous settlement (see Van den Assem 1971), and is predicted to lead to high densities, as rules of precedence over access to territories will not apply. Arbitrary patterns of sequential establishment lead to distances between nearest neighbours which are greater than those between neighbours under simultaneous settlement (for simple models, see Maynard Smith 1974; Tanemura and Hasegawa 1980). Taitt argues that a major irruption often occurs when there is no spring decline in densities, and that the absence of a spring decline in years of plenty causes dumping of females into the population and may precipitate an irruption. The first experimental attempt to test this hypothesis in *Microtus townsendii* was not successful (Taitt 1985b). However, important components of the model can be verified in microtine populations. First, the presence of surplus animals which might compete for territories has been demonstrated repeatedly, in an attempt to demonstrate Watson and Moss's (1970) conditions for social regulation of populations (e.g. Krebs *et al.* 1976). Second, Baird and Birney (1982) argue that the presence of a population of *M. pennsylvanicus* is most effective in restricting settlement when animals have been resident for a long period of time. Third, invasion into unoccupied habitat is a conspicuous component of population irruptions (Chapter 2), as is a relaxation of stringency in the non-breeding season preceding the irruption (Cockburn and Lidicker 1983; Hansson 1984b; Kaikusalo and Tast 1984, and references therein).

In an interesting elaboration of Maynard Smith's (1974) model, Getty (1981) has contended that the difference between the consequences of sequential and simultaneous settlement promotes 'collusion' between animals establishing territories early in a sequence, in order to preclude later establishment of rivals and subsequent compression of territories. Animals behaving according to his predictions should not establish territories at random, but should cluster around early 'seed' territories. Unlike random settlement, nearest neighbour distances should not decline as the habitat is saturated. The consequence is a high mean territory size but a low variance. Within microtines, all the conditions necessary for Getty's hypothesis of preemption are clearly satisfied, though there is still uncertainty about whether or not animals settle in the first empty site they encounter (compare Tamarin 1977a with Baird and Birney 1982). The relation between these observations and clustering of female voles in some species requires elaboration.

Territoriality.
It should be clear from the simultaneous settlement hypothesis that the population consequences of territoriality and those of dispersal are closely related. It should also be apparent that territoriality will limit the number of animals which can settle within an area. One interesting body of theory was initially developed by Lomnicki (1978, 1980) and Gurney and Nisbet (1979). They show that where resources are distributed unevenly among the population (into 'haves and have-nots') the conditions for stability are extended greatly. These analyses refer to the case where some territories provide much better resources than others. It may pay some individuals to emigrate from such a population, even when dispersers suffer high mortality (enter a dispersal sink) (Lomnicki 1980). This contributes further to stability (Lomnicki 1982). In a recent analysis, Lomnicki and Ombach (1984) have asked what is the chief factor contributing to population stability - the unequal resource partitioning, or the ability of animals with access to good quality resources to insulate themselves from the deleterious effects of density? The latter effect proved to be most important, and hence that application to mammalian data is dependent on their exhibiting territoriality.

It is also worth recollecting the work which suggests that territory size in some vertebrate species is a reflection of the probability of future resource shortage, and hence is determined more directly by intruder pressure than by food availability (Lima 1984). Increases in territory size of this sort will contribute to stability.

Summary.
We have excellent evidence consistent with theoretical predictions that although dispersal contributes to the expansion of populations from constant habitat, it dampens the rate of population increase within that habitat, and that territoriality limits the density of breeding animals. But there is *no* evidence that either factor changes in a way which could precipitate a decline, or that changes in either behaviour precedes a population increase. Neither appear to be manifested more strongly in the peak and decline phase of population fluctuations (Stenseth 1983; Chapter 3). In view of the intense focus on this hypothesis, this failure to produce any confirmation is particularly damning.

Seasonality

I have already made clear my belief that the microtine life history is a reflection of the evolutionary pressures associated with living in a highly seasonal environment. The consequences of seasonality in population models has attracted remarkably little attention, yet the existence of seasonal hiatuses in reproduction run counter to all simple depictions of population growth (e.g. the logistic). Kot and Schaffer (1984) used a series of simple difference equations to simulate the effects of varying seasonality in habitats with different productivity. Because these models are based on difference equations, these models are most appropriate for bivoltine insects, but the results are nonetheless of great interest for microtine biologists. This is particularly true given the success of the same authors in rendering population sequences of vertebrates into difference equations.

In less productive environments, seasonality is destabilising. As seasonality increases, the dynamics move from higher order multiannual cycles to chaos. However, in more productive environments, a small amount of seasonality can generate greater stability than either an absence of seasonality or extreme values. More than one result exists under some conditions of seasonality and productivity, though this complexity is never revealed in analysis of the simple logistic equation.

The other condition of interest was resolved by graphical analysis. It concerns the time of year in which the population is regulated (winter or summer) and determines the extent to which oscillations will be in phase or out of phase with environmental fluctuations. Out-of-phase cycles occur under summer regulation and are predicted in high productivity, low seasonality environments. If the difference between summer and winter is very great, Kot and Schaffer (1984) predict that post-summer crashes may be very severe and several seasons may be required for recovery. Hence in productive environments, asynchronous trajectories should develop with increasing distance from the equator. In less productive habitats, asynchronous cycles of increasing magnitude should develop with increasing latitude. There is little doubt that these conclusions effectively predict the behaviour of microtine populations, though the formal mathematics for populations of mammals with overlapping generations have not been developed.

In summary, it does appear likely that seasonality can contribute to cyclic dynamics under certain conditions, but appropriate data for microtines need to be gathered.

Relaxation of Seasonality

If the season of stringency is implicated in the evolution of life histories, any relaxation in stringency is of interest as a source of unusual demographic fluctuations. With this in mind, much attention has been paid to winter breeding by microtines. There seems little doubt that food often plays a role in the stimulation of winter breeding (Andrzejewski 1975; Hickie, Lavigne and Woodward 1982; Eriksson 1984; Kaikasulo and Tast 1984), but this trend is not universal (Hansson 1984b). Improved winter nutrition also appears to enhance reproduction in *Lepus timidus* (Pehrson and Lindlöf 1984), and *Lepus americanus* (Keith and Windberg 1978). Direct correlations between improved conditions in the season of stringency, and both winter breeding and population outbreaks, have been documented for many populations (e.g. Hamilton 1941; Krebs 1964a; Keller and Krebs 1970; Myllymäki 1977a; Smolen and Keller 1979; Cockburn and Lidicker 1983). Jannett (1984) and Hansson (1984b) provide a comprehensive review of North American and Scandinavian data respectively. Hansson (1984b) points out that winter breeding is more prevalent at northern latitudes.

While winter reproduction probably makes only a small demographic contribution (Hansson 1984b), it may be indicative of amelioration of conditions and an increase in the source population from which the spring increase will be derived. If more habitat is available in the period preceding the spring increase, the expansion of voles into colonisation habitat may contribute to the extent of increase, both on a microgeographic and macrogeographic scale. It therefore seems possible that changes to the usual pattern of seasonal adversity may contribute to population increase, with ordinary adversity depressing numbers in the following season. But food supplementation appears unable to defer declines indefinitely, so the relevance of these factors to all aspects of cyclic phenomena remains elusive.

Population Dynamics and Social Behaviour

In Chapter 1, I identified three questions which have dominated recent research on fluctuating populations of vertebrates. It is possible in the light of the discussion to reconsider these questions and see if my review of the literature has shed any light on these questions.

The Problem of Amplitude

I believe that both theory and empirical evidence now support the idea that behavioural dynamics in fluctuating populations of vertebrates contribute principally towards restriction of population increase, and consequently drive populations towards stability, rather than towards cyclicity (or chaos). While there is evidence that changes in spacing behaviour can precipitate temporary declines during the spring (Taitt and Krebs 1985), there is no evidence that equivalent effects generate the dramatic crashes which occasionally follow a peak year. Hypotheses which related population dynamics directly to nutrition foundered on the inability of food supplementation to prevent a decline. While the critical experiments to test hypotheses of social organisation await the solution to methodological problems, I confess a lack of optimism that they will fare any better.

The Problem of Cyclicity

I have already pointed out that high densities may be unnecessary for cyclicity. Instead, it may be an endogenous property of populations reflecting the period required to reproduce, and hence to precipitate population increase. Populations with high fecundity exhibit dynamics of high periodicity which generate a complex spectrum of 'waves' in age structure (Oster and Takashi 1974), and this may lead to synchronised multiannual fluctuations (see also Finerty 1980). Synchrony in patterns of reproduction should be facilitated where the population is reduced by extrinsic factors to very low levels. This bottlenecking may be most severe in environments characterised by extremes of seasonality. The mathematics of these relations remain unexplored, but such interactions represent a promising lead for future theoretical work.

The Problem of the Changing Social Environment - the Chitty Hypothesis Revisited

Although promising data are available for variation in patterns of reproduction, the case that aspects of social evolution are unique to cyclic populations is difficult to sustain. This is particularly true for properties such as territoriality and dispersal, to which recent population theory attributes great importance. Earlier work on variation in aggression is very difficult to interpret, as conflicting results have been obtained for the taxa of interest. For example, grouse become increasingly aggressive during population declines (Moss, Watson and Rothery 1984), while voles and snowshoe hares appear to be less aggressive in declines than in peak populations, though data are weak (De Poorter 1984; C.J. Krebs 1985b). Neither data support the Chitty Polymorphic Behaviour Hypothesis. Can we interpret these changes in terms of selection at all? As a tentative first step, I offer the following.

In Chapter 1, I argued that the Chitty Polymorphic Behaviour Hypothesis had been falsified both theoretically and empirically, but represented a major contribution because it focused attention on the possibility that natural selection operates on an ecological time scale. This conclusion has been vindicated by some of the studies of seasonal selection in house mice and finches. I wish to conclude by a brief speculative account concerning the advantages of aggressive behaviour at different stages of a population fluctuation. While I am convinced that changes of this sort contribute very little to dramatic population oscillations, the problem of density-dependence in aggressive behaviour does warrant attention.

The evidence from contrived encounters that aggression in voles is only weakly developed in the decline phase of a microtine population (C.J. Krebs 1985b), is not only in direct contrast to the central prediction of the original and derivative versions of the Chitty hypothesis, but also runs counter to the almost ubiquitous view that interference mechanisms will be most pronounced when density is high or declining (Gill 1974). But consider the following. In contrast to the panmixia assumed in naive population theory, vertebrate populations are highly structured. Philopatry, even if confined to one sex, means that interactions between members of that sex are non-random with respect to relatedness. In a population which is expanding rapidly there must be a large number of vacant sites. Because seasonal and successional effects will enforce synchrony in the availability of sites to dispersers, there may be simultaneous settlement, as proposed by Taitt (1985b). Under these circumstances, aggression may be highly advantageous, occuring principally between unrelated dispersers. As the availability of vacant sites decreases, the prospects of location of vacant sites will decline. Rather than escape in space (dispersal), escape in time (diapause) may be selected, generating the variation in maturation schedules I documented in Chapter 5. Because immigration will decline, and there is a reasonably high level of philopatry, animals will be increasingly surrounded by relatives which have dispersed over short distances, or which have failed to disperse at all (e.g Jannett 1978). If there is heritability of aggression, 'aggressive genotypes' will interact mainly with other similarly aggressive animals, and will sustain the high frequency-dependent costs of aggression that result from the consequent escalated contests (see Maynard Smith and Price 1973; Maynard Smith 1982). Note the difference between this hypothesis and the 'kin selection' hypothesis of Charnov and Finerty (1980), which was developed by Stenseth (1983) and Moss and Watson (1985). Under the combined conditions of frequency- and density-dependence, aggressive family groups will perform less well than tolerant animals, and their frequency should intially decline, and then oscillate according to local circumstances. According to this hypothesis, aggression should be selected throughout population increase, but should be strongly selected against in peak conditions, leading to low frequencies during a decline. To summarise, I believe the correlation between population expansion and aggressive behaviour has been interpreted without regard to frequency-dependence and population structure, and may have consequently generated conclusions in the wrong direction.

But what of the contrary evidence for grouse populations, where aggressiveness continues to increase as the decline persists? Unlike microtines, the vagility of grouse and ptarmigan enables them to depart their breeding ranges, and the males establish territories which determine their capacity to attract mates. Male removal after birds are paired causes the departure of the hen (Watson and Jenkins 1968), though hens will settle in the absence of males (Hannon 1983, 1984). The importance of males in determining the pattern of settlement and recruitment is reflected in the tendency of males rather than females to be philopatric. This is in sharp contrast to the overwhelming evidence that females are the principal influence on recruitment in microtine populations. Although this difference offers some promise, it is clear that the inconsistency between the behaviour of different taxa which exhibit cyclic fluctuations in numbers reduces the prospects for a simple synthetic theory of social regulation. Nonetheless, polymorphism in aggressive behaviour may yet prove to be of considerable interest, and attention to the interspecific and intraspecific variation a rewarding field of study.

Conclusions

Voles and lemmings and their peculiar habits have influenced a generation of population ecologists. May they continue to do so. If this book has a general theme, it is that we must beware interpreting the ecology of a group of animals in terms of the single factor which to us is most conspicuous - in this case, density. My discussion should also illustrate that ecology is a synthetic science. Just as biologists fascinated by population regulation in voles should pay heed to the lessons afforded by those more tractable species, whose populations are predictable and behaviour more easily studied, behavioural ecologists can learn many lessons from these remarkable little creatures. I hope this account will encourage both audiences towards this union.

References

Ågren, G. 1984. Incest avoidance and bonding between siblings in gerbils. *Behav. Ecol. Sociobiol. 14* : 161-9
_____ 1985. Alternative mating strategies in the Mongolian gerbil. *Behaviour 91* : 229-44
Alcock, J. 1984. *Animal Behavior: An Evolutionary Approach.* 3rd edn. Sinauer, Sunderland, Massachusetts
Alder, E.M., Godfrey, J., McGill, T.E. and Watt, K.R. 1981. The contributions of genotype and sex to variation in mating behaviour between geographical subspecies of the bank vole (*Clethrionomys glareolus* Schreber). *Anim. Behav. 29* : 942-52
Alerstam, T. and Högstedt, G. 1984. How important is clutch size dependent mortality? *Oikos 43* : 253-4
Andersen, T. and Wiig, Ø. 1982. Epigenetic variation in a fluctuating population of lemming (*Lemmus lemmus*) in Norway. *J. Zool. Lond. 197* : 391-404
Anderson, J.H. 1975. Phenotypic correlations among relatives and variability in reproductive performance in populations of the vole *Microtus townsendii*. Unpublished Ph.D. thesis, University of British Columbia
Anderson, P.K. 1970. Ecological structure and gene flow in small mammals. *Symp. Zool. Soc. Lond. 26* : 299-325
_____ 1980. Evolutionary implications of microtine behavioral systems on the ecological stage. *The Biologist 62* : 70-88
_____, Whitney, P.H. and Huang, J.-P. 1976. *Arvicola richardsoni*: ecology and biochemical polymorphism in the front ranges of southern Alberta. *Acta theriol. 21* : 425-68
Anderson, R.M. and May, R.M. 1978. Regulation and stability of host-parasite population interactions. I. Regulatory processes. *J. Anim. Ecol. 47* : 219-47
Andersson, M. and Erlinge, S. 1977. Influence of predation on rodent populations. *Oikos 29* : 591-7
_____ and Jonasson, S. 1986. Rodent cycles in relation to food resources on alpine heath. *Oikos 46* : 93-106
Andrzejewski, A. 1975. Supplementary food and winter dynamics of bank vole populations. *Acta Theriol 20* : 23-40
Angelstam, P., Lindström, E. and Widén, P. 1985. Synchronous short-term population fluctuations of some birds and mammals in Fennoscandia - occurrence and distribution. *Holarctic Ecology 8* : 285-98
Archer, S. and Tieszen, L.L. 1983. Effects of simulated grazing on foliage and root production and biomass allocation in an arctic tundra sedge (*Eriophorum vaginatum*). *Oecologia (Berl.) 58* : 92-102
Ashmole, N.P. 1963. The regulation of numbers of tropical oceanic birds. *Ibis 103b* : 458-73
Atchley, W.R. 1984. Ontogeny, timing of development, and genetic variance-covariance structure. *Amer. Natur. 123* : 519-40
_____ and Rutledge, J.J. 1980. Genetic components of size and shape. I. Dynamics of components of phenotypic variability and covariability during ontogeny in the laboratory rat. *Evolution 34* : 1161-73
Baddaloo, E.G.Y. and Clulow, F.V. 1981. Effects of the male on growth, sexual

maturation, and ovulation of young female meadow voles, *Microtus pennsylvanicus*. *Can. J. Zool. 59* : 415-21

Baird, D.D. and Birney, E.C. 1982. Pattern of colonization in *Microtus pennsylvanicus*. *J. Mamm. 63* : 290-3

Banks, E.M., Brooks, R.J. and Schnell, J. 1975. A radiotracking study of home range and activity of the brown lemming (*Lemmus trimucronatus*). *J. Mamm. 56* : 888-901

_____, Mankovich, N.J. and Huck, U.W. 1979. Female interspecific aggression in two species of lemmings. *Behav. Neur. Biol. 26* : 372-8

Barnett, S.A. and Dickson, R.G. 1985. A paternal influence on survival of wild mice in the nest. *Nature 317* : 617-8

Barrowclough, G.F. 1978. Sampling bias in dispersal studies based on finite area. *Bird banding 49* : 333-41

Barry, R.E. 1984. Effects of early experience on cue preferences of *Peromyscus leucopus*. *Amer. Midl. Natur. 111* : 234-41

Barton, N.H. 1987. The genetic consequences of dispersal. In N.C. Stenseth and W.Z. Lidicker (eds.) *Animal Dispersal : Small Mammals as a Model*. Chapman & Hall, London, pp. 000-000

Bateman, A.J. 1948. Intra-sexual selection in *Drosophila*. *Heredity 2* : 349-68

Bateson, P. 1982. Preference for cousins in Japanese quail. *Nature 295* : 236-7

_____ 1983. Optimal outbreeding. In P. Bateson (ed.) *Mate Choice*. Cambridge University Press, Cambridge, pp. 257-77

_____, Lowick, W. and Scott, D.K. 1980. Similarities between the faces of parents and offspring in Bewick's swan and the differences between mates. *J. Zool. Lond. 191* : 61-74

Batzli, G.O. 1975. The role of small mammals in arctic ecosystems. In K. Petrusewicz, F.B. Golley and L. Ryszkowski (eds.) *Small Mammals: Productivity and Dynamics of Populations*. Cambridge University Press, Cambridge, pp. 243-68

_____ 1985. Nutrition. In R.H. Tamarin (ed.), *Biology of New World* Microtus, Special Publication No. 8 , American Society of Mammalogists, pp. 779-811

_____ and Cole, F.R. 1979. Nutritional ecology of microtine rodents : digestibility of forage. *J. Mamm. 60* : 740-50

_____ and Jung, H.G. 1980. Nutritional ecology of microtine rodents : resource utilization near Atkasook, Alaska. *Arctic Alpine Res. 12* : 483-99

_____ , Getz, L.L. and Hurley, S.S. 1977. Suppression of growth and reproduction of microtine rodents by social factors. *J. Mamm. 58* : 583-91

_____ and Pitelka, F.A. 1971. Condition and diet of cycling populations of the California vole, *Microtus californicus*. *J. Mamm. 52* : 141-63

_____ and _____ 1975. Vole cycles : test of another hypothesis. *Amer. Natur 109* : 482-7

_____ and _____ 1983. Nutritional ecology of microtine rodents : food habits of lemmings near Barrow, Alaska. *J. Mamm. 64* : 648-55

_____ , _____ and Cameron, G.N. 1983. Habitat use by lemmings near Barrow, Alaska. *Holarctic Ecol. 6* : 255-62

_____ , White, R.G., McClean, S.F., Pitelka, F.A. and Collier, B. 1980. The herbivore-based trophic system. In J.Brown et al. (eds.) *An Arctic Ecosystem : The Coastal Plain of Northern Alaska*. Dowden, Hutchinson and Ross, Stroudsburg, Pennslyvania, pp. 355-410

Beacham, T.D. 1979a. Dispersal tendency and duration of life of littermates during population fluctuations of the vole *Microtus townsendii*. *Oecologia (Berl.) 42* : 11-22

_____ 1979b. Size and growth characteristics of dispersing voles, *Microtus townsendii*. *Oecologia (Berl.). 42* : 1-10

_____ 1980a. Dispersal during population fluctuations of the vole, *Microtus townsendii*. *J. Anim. Ecol. 49* : 867-77

_____ 1980b. Growth rates of the vole *Microtus townsendii* during a population cycle. *Oikos 35* : 99-106

_____ 1980c. Growth rates of aggressive and docile voles, *Microtus townsendii*. *Amer. Midl. Natur. 104* : 387-89

_____ 1981. Some demographic aspects of dispersers in fluctuating populations of the vole *Microtus townsendii*. *Oikos 36* : 273-80

_____ and Krebs, C.J. 1980. Pitfall versus live-trap enumeration of fluctuating populations of *Microtus townsendii*. *J. Mamm. 61* : 486-99

Beauchamp, G.K., Yamazaki, K. and Boyse, E.A. 1985. The chemosensory recognition of genetic individuality. *Scient. Amer. 253(1)* : 66-73

Begon, M. 1984. Density and individual fitness : asymmetric competition. In B. Shorrocks (ed.) *Evolutionary Ecology*. Blackwell Scientific Publications, Oxford, pp. 175-94

_____ 1985. A general theory of life-history variation. in R.M. Sibly and R.H. Smith (eds.) *Behavioural Ecology: Ecological Consequences of Adaptive Behaviour*. Blackwell Scientific Publications, Oxford, pp. 91-7

_____ and Mortimer, M. 1981. *Population Ecology : a Unified Study of Plants and Animals*. Blackwell Scientific Publications, Oxford

Beilharz, R.G. 1975. The aggressive response of male mice (*Mus musculus* L.) to a variety of stimulus animals. *Z. Tierpsychol. 39* : 141-9

Bell, D.J. and Mitchell, S. 1984. Effect of female urine on growth and sexual maturation in male rabbits. *J. Reprod. Fert. 71* : 155-60

_____ and Reece, C. 1983. An investigation into the "Bruce effect" in domesticated rabbits. In D. Muller-Schwarze & R.M. Silverstein (eds.) *Chemical Signals in Vertebrates. Vol. 3*. Plenum Press, New York, pp. 339-42

Bell, G. 1984a. Measuring the cost of reproduction. I. The correlation structure of the life table of a planktonic rotifer. *Evolution 38* : 300-13

_____ 1984b. Measuring the cost of reproduction. II.The correlation structure of the life tables of five freshwater invertebrates. *Evolution 38* : 314-26

Belovsky, G.E. 1984. Snowshoe hare optimal foraging and its implications for population dynamics. *Theor. Popul. Biol. 22* : 235-64

Benenson, I.E. 1983. On the maintenance of the unique system of sex determination in lemmings. *Oikos 41* : 211-8

Bengtsson, B.O. 1977. Evolution of the sex ratio in the wood lemming. In F.B. Christiansen & T.M. Fenchel (eds.) *Measuring Selection in Natural Populations*. Springer-Verlag, Berlin, pp. 333-43

_____ 1978. Avoid inbreeding: at what cost? *J. theor. Biol. 73* : 439-44

Berger, P.J. and Negus, N.C. 1982. Stud male maintenance of pregnancy in *Microtus montanus*. *J. Mamm. 63* : 148-51

_____ , _____ , Sanders, E.H. and Gardner, P.D. 1981. Chemical triggering of reproduction in *Microtus montanus*. *Science 214* : 69-70

Bergeron, J.-M. 1980. Importance des plantes toxiques dans le régime alimentaire de *Microtus pennsylvanicus* à deux étapes opposées de leur cycle. *Can. J. Zool. 58* : 2230-8

_____ 1984. L'utilisation de resources alimentaires avec or sans composés secondaires connus par une population cyclique de campagnols des champs (*Microtus pennsylvanicus*). *Can. J. Zool. 62* : 601-7

Bergerud, A.T. and Mossop, D.H. 1984. The pair bond in ptarmigan. *Can. J. Zool. 62* : 2129-41

Berry, R.J. 1970. Covert and overt variation, as exemplified by British mouse

192 REFERENCES

populations. *Symp. zool. Soc. Lond* 26 : 3-26
_____ 1977. The population genetics of the house mouse. *Sci. Prog., Oxf.* 64 : 341-70
_____ 1981. Population dynamics of the house mouse. *Symp. Zool. Soc. Lond.* 47 : 395-425
_____ and Jakobson, M.E. 1971. Life and death in an island population of the house mouse. *Exp. Geront.* 6 : 187-97
Berven, K.A. and Gill, D.E. 1983. Interpreting geographic variation in life-history traits. *Amer. Zool.* 23 : 85-97
Birney, E.C., Grant, W.E. and Baird, D.D. 1976. Importance of vegetative cover to cycles of *Microtus* populations. *Ecology* 57 : 1043-51
Blaustein, A.R. 1980. Behavioural aspects of competition in a three-species rodent guild in coastal southern California. *Behav. Ecol. Sociobiol.* 6 : 247-55
_____ 1981. Population fluctuations and extinctions of small rodents in coastal southern California. *Oecologia (Berl.)* 48 : 71-8
_____ 1983. Kin recognition mechanisms : phenotypic matching or recognition alleles. *Amer. Natur.* 121 : 749-54
_____ and O'Hara, R.K. 1985. Kin recognition in tadpoles. *Scient. Amer.* 254 : 90-96
Boag, P.T. and Grant, P.R. 1981. Intense natural selection in a population of Darwin's finches (Geospizinae) in the Galápagos. *Science* 214 : 82-5
_____ and Van Nordwijk, A.J. 1987. The quantitative genetics of wild bird populations. In F. Cooke & P.A. Buckley (eds.) *Avian Genetics : Its Application to Ecological and Evolutionary Theory.* Academic Press, New York, Pp. 000-000
Bodmer, W.F., Trowsdale, J., Young, J. and Bodmer, J. 1986. Gene clusters and the evolution of the major histocompatibility system. *Phil. Trans. R. Soc. Lond.* B312 : 303-15
Bondrup-Nielsen, S. 1985. Ecology of the wood lemming, *Myopus schisticolor*. In W.A. Fuller, M.T. Nietfeld & M.A. Harris (eds.) *Abstracts of Papers and Posters, Fourth International Theriological Congress.* Edmonton, 13-20 August 1985 : Edmonton
_____ 1986. Investigation of spacing behaviour of *Clethrionomys gapperi* by experimentation. *J. Anim. Ecol.* 55 : 269-80
_____ and Ims, R.A. 1986. Reproduction and spacing behaviour in females in a peak density population of *Clethrionomys glareolus*. *Holarctic Ecology* 9 : 109-12
_____ and Karlsson, F. 1985. Movements and spatial patterns in populations of *Clethrionomys* species`: a review. *Ann. Zool. Fennici* 22 : 385-92
Boonstra, R. 1977. Effects of conspecifics on survival during population declines in *Microtus townsendii*. *J. Anim. Ecol.* 46 : 835-51
_____ 1978. Effect of adult townsend voles (*Microtus townsendii*) on survival of young. *Ecology* 59 : 242-8
_____ 1984. Aggressive behavior of adult meadow voles (*Microtus pennsylvanicus*) towards young. *Oecologia (Berl.)* 62 : 126-31
_____ and Krebs, C.J. 1977. A fencing experiment on a high-density population of *Microtus townsendii*. *Can. J. Zool.* 55 : 1166-75
_____ and _____ 1978. Pitfall trapping of Microtus *townsendii*. *J. Mamm.* 59 : 136-48
_____ and _____ 1979. Viability of large- and small-sized adults in fluctuating vole populations. *Ecology* 60 : 567-73
_____ and Rodd, F.H. 1983. Regulation of breeding density in *Microtus pennsylvanicus*. *J. Anim. Ecol.* 52 : 757-80
_____ and _____ 1984. Efficiency of pitfalls versus live traps in enumeration of populations of *Microtus pennsylvanicus*. *Can. J. Zool.* 62 : 758-65
_____ and Youson, J. 1982. Hip glands in a field population of *Microtus*

pennsylvanicus. Can. J. Zool. 60 : 2155-958

Boström, U. and Hansson, L. 1981. Small rodent communities on mires : implications for population performance in other habitats. *Oikos* 37 : 216-24

Boutin, S. 1980. Effect of spring removal experiments on the spacing behaviour of female snowshoe hares. *Can. J. Zool.* 58 : 2167-74

―――― 1984a. Effect of late winter food addition on numbers and movements of snowshoe hares. *Oecologia (Berl.)* 62 : 393-400

―――― 1984b. The effects of conspecifics on juvenile survival and recruitment of snowshoe hares. *J. Anim. Ecol.* 53 : 623-37

――――, Gilbert, B.S., Krebs, C.J., Sinclair, A.R.E. and Smith, J.N.M. 1985. The role of dispersal in the population dynamics of snowshoe hares. *Can.J. Zool.* 63 : 106-15

Bowen, B.S. 1982. Temporal dynamics of microgeographic structure of genetic variation in *Microtus californicus. J. Mamm.* 63 : 625-38

Bowen, D.W. and Brooks, R.J. 1978. Social organization of confined male collared lemmings (*Dicrostonyx groenlandicus* Traill). *Anim. Behav.* 26 : 1126-35

Boyce, C. and Boyce, J.L. 1985. Groups of breeding females in Microtus arvalis. In W.A. Fuller, M.T. Nietfeld & M.A. Harris (eds.) *Abstracts of Papers and Posters, Fourth International Theriological Congress.* Edmonton, 13-20 August 1985, Edmonton

Boyce, M.S. 1979. Seasonality and patterns of natural selection for life histories. *Amer. Natur.* 114 : 569-83

―――― 1984. Restitution of r- and K-selection as a model of density-dependent natural selection. *Ann. Rev. Ecol. Syst.* 15 : 427-47

―――― and Miller, R.S. 1985. Ten-year periodicity in whooping crane census. *Auk* 102 : 658-60

Boyd, S.K. and Blaustein, A.R. 1985. Familiarity and inbreeding avoidance in the gray-tailed vole (*Microtus canicaudus*). *J. Mamm.* 66 : 348-52

Bradbury, J.W., Vehrencamp, S.L. and Gibson, R. 1985. Leks and the unanimity of female choice. In P.J. Greenwood, P.H. Harvey & M. Slatkin (eds.) *Evolution: Essays in Honour of John Maynard Smith.* Cambridge University Press, Cambridge, pp. 301-14

Braithwaite, R.W. and Lee, A.K. 1979. A mammalian example of semelparity. *Amer. Natur.* 113 : 151-6

Brody, S. 1945. Time relations of growth of individual populations. In *Bioenergetics and Growth.* Reinhold, New York, pp. 487-574.

Bronson, F.H. 1979. The reproductive ecology of the house mouse. *Q. Rev. Biol.* 54 : 265-99

Brooks, R.J. 1984. Causes and consequences of infanticide in populations of rodents. In G. Hausfater & S.B. Hrdy (eds.) *Infanticide : Comparative and Evolutionary Perspectives.* Aldine Publishing Company, New York. pp. 331-48

―――― and Schwarzkopf, L. 1983. Factors affecting incidence of infanticide and discrimination of related and unrelated neonates in male *Mus musculus. Behav. Neur. Biol.* 37 : 149-61

―――― and Webster, A.B. 1984. Relationship of seasonal changes in age structure and body size in *Microtus pennsylvanicus.* In J.F. Merritt (ed.), *Winter Ecology of Small Mammals.* Special Publication No. 10 : Carnegie Museum of Natural History, Pittsburgh, pp. 275-84

Brown, E.R. 1973. Changes in seasonal patterns of growth in *Microtus pennsylvanicus. Ecology* 54 : 1103-10

Brown, J.L. 1964. The evolution of diversity in avian territorial systems. *Wilson Bulletin* 76 : 160-9

―――― 1982. Optimal group size in territorial animals. *J. theor. Biol.* 95 : 793-810

Bruce, H.M. 1959. An exteroceptive block to pregnancy in the mouse. *Nature* 184 : 105

_____ 1960. A block to pregnancy in the mouse caused by the proximity of strange males. *J. Reprod. Fertil. 1* : 96-103

_____. and Parrott, D.M.V. 1960. Role of olfactory sense in pregnancy block by strange males. *Science 131* : 1526

Bryant, J.P. 1981. Phytochemical deterrence of snowshoe hare browsing by adventitious shoots of four Alaskan trees. *Science 213* : 889-90

_____ and Kuropat, P. 1980. Selection of winter forage by subarctic browsing vertebrates : the role of plant chemistry. *Ann. Rev. Ecol. Syst. 11* : 261-85

_____, Wieland, G.D., Reichardt, P.B., Lewis, V.E. and McCarthy, M.C. 1983. Pinosylvin methyl ether deters snowshoe hare feeding on green alder. *Science 222* : 1023-5

_____, _____, Clausen, T. and Kuropat, P. 1985. Interactions of snowshoe hare and feltleaf willow in Alaska. *Ecology 66* : 1564-73

Bujalska, G. 1973. The role of spacing behaviour among females in the regulation of reproduction in the bank vole. *J. Reprod. Fertil. Suppl. 19* : 465-754

_____ 1985a. Regulation of female maturation in *Clethrionomys* spp., with special reference to an island population of *C. glareolus*. *Ann. Zool. Fennici 22* : 385-92

_____ 1985b. Fluctuations in an island bank vole population in the light of the study on its organization. *Acta theriol. 30* : 1-49

Bull, J.J. 1983. *Evolution of Sex Determining Mechanisms*. Benjamin/Cummings, Menlo Park, California

_____ and Bulmer, M.G. 1981. The evolution of XY females in mammals. *Heredity 39* : 1-14

Bulmer, M.G. 1984. Risk avoidance and nesting strategies. *J. theor. Biol. 106* : 529-35

_____ and Taylor, P.D. 1980a. Sex ratio under the haystack model. *J. theor. Biol. 86* : 83-9

_____ and _____ 1980b. Dispersal and the sex ratio. *Nature 284* : 448-9

Bumpus, H.C. 1899. The elimination of the unfit as illustrated by the introduced sparrow, *Passer domesticus*. *Biol. Lectures, Marine Biol. Lab., Woods Hole, MA.*, pp. 209-26

Burns, G.R. 1981. Population dynamics of island populations of subarctic *Clethrionomys rutilus*. *Can. J. Zool. 59* : 2115-22

Bygott, J.D., Bertram, B.C.R. and Hanby, J. 1979. Male lions in large coalitions gain reproductive advantages. *Nature 282* : 839-41

Calder, W.A. 1983. An allometric approach to population cycles of mammals. *J. theor. Biol. 100* : 275-82

_____ 1984. *Size, Function and Life History*. Harvard University Press, Cambridge, Massachusetts.

Caley, J. and Boutin, S. 1985. Infanticide in wild populations of *Ondatra zibethicus* and *Microtus pennsylvanicus*. *Anim. Behav. 33* : 1036-7

Calhoun, J.B. 1962. Population density and social pathology. *Sci. Am. 206* : 139-48

Cannings, C. and Thompson, E.A. 1981. *Genealogical and Genetic Structure*. Cambridge University Press, Cambridge

Caraco, T., Martindale, S. and Whitham, T.S. 1980. An empirical demonstration of risk-sensitive foraging preferences. *Anim. Behav. 28* : 820-30

Carothers, A.D. 1980. Population dynamics and the evolution of sex-determination in lemmings. *Genet. Res. 36* : 199-209

Carstairs, J.L. 1976. Population dynamics and movements of *Rattus villosissimus* (Waite) during the 1966-69 plague at Brunette Downs, Northern Territory. *Aust. Wildl. Res. 3* : 1-10

Carter, C.S., Getz, L.L., Gavish, L., McDermott, J.L. and Arnold, P. 1980. Male-related pheromones and the activation of female reproduction in the prairie vole (*Microtus ochrogaster*). *Biol. Reprod. 23* : 1038-45

Cary, J.R. and Keith, L.B. 1979. Reproductive change in the 10-year cycle of snowshoe hares. *Can. J. Zool.* 57 : 375-90

Caryl, P.G. 1979. Communication by agonistic displays: what can games theory contribute to ethology? *Behaviour 68* : 136-69

Castro-Sierra, E. and Wolf, U. 1967. Replication patterns of the chromosome No. 9 of the rodent *Ellobius lutescens*. *Cytogenetics 6* : 268-75

Caswell, H. 1982. Life history theory and the equilibrium status of populations. *Amer. Natur. 120* : 317-39

―――― 1983. Phenotypic plasticity in life-history traits : demographic effects and evolutionary consequences. *Amer. Zool. 23* : 35-46

Cates, R.G. and Orians, G.H. 1975. Successional status and the palatibility of plants to generalized herbivores. *Ecology 56* : 410-18

Cengel, D.J., Estep, J.E. and Kirkpatrick, R.L. 1978. Pine vole reproduction in relation to food habits and body fat. *J. Wildl. Mgmt. 42* : 822-33

Charlesworth, B. 1971. Selection in density regulated populations. *Ecology 52* : 469-74

―――― 1980. *Evolution in Age-Structured Populations*. Cambridge University Press, Cambridge

―――― 1984. The evolutionary genetics of life histories. In B. Shorrocks (ed.) *Evolutionary Ecology*. Blackwell Scientific Publications, Oxford, pp. 117-33

―――― and Giesel, J.T. 1972 Selection in populations with overlapping generations. II. Relations between gene frequency and demographic variables. *Amer. Natur. 106* : 388-401

Charnov, E.L. 1982. *The Theory of Sex Allocation*. Princeton University Press, Princeton

―――― and Finerty, J.P. 1980. Vole population cycles. A case for kin-selection? *Oecologia (Berl.) 45* : 1-2

―――― and Krebs, J.R. 1974. On clutch-size and fitness. *Ibis 116* : 217-9

Cheverud, J.M., Dow, M.M. and Leutenegger, W. 1985. The quantitative assessment of phylogenetic constraints in comparative analyses : sexual dimorphism in body weight among primates. *Evolution 39* : 1335-51

――――, Rutledge, J.J. and Atchley, W.R. 1983. Quantitative genetics of development: Genetic correlations among age-specific trait values and the evolution of ontogeny. *Evolution 37* : 895-905

Chitty, D. 1952. Mortality among voles (*Microtus agrestis*) at Lake Vyrnwy, Montgomeryshire in 1936-9. *Phil. Trans. Roy. Soc. Lond. B36* : 505-52

―――― 1958. Self-regulation of numbers through changes in viability. *Cold Spring Harbor Symp. Quant. Biol. 22* : 277-80

―――― 1959. A note on shock disease. *Ecology 40* : 728-31

―――― 1960. Population processes in the vole and their relevance to general theory. *Can. J. Zool. 38* : 99-113

―――― 1967. The natural selection of self-regulatory behaviour in animal populations. *Proc. Ecol. Soc. Aust. 2* : 51-78

Chitty, H. and Chitty, D. 1962. Body weight in relation to population phase in Microtus agrestis. In J. Kratchovil and J. Pelikan (eds.) *Symposium Theriologicum: Proceedings of the International Symposium on Methods of Mammalogical Investigation*. Czechoslovak Academy of Sciences, Prague, pp. 77-86

Christian, J.J. 1950. The adreno-pituitary system and population cycles in small mammals. *J. Mamm. 31* : 247-59

―――― 1961. Phenomena associated with population density. *Proc. Natn. Acad. Sci USA 47* : 428-49

―――― 1971. Population density and reproductive efficiency. *Biol. Reprod. 4* : 248-94

―――― 1975. Hormonal control of population growth. In B.E. Eleftheriou and R.L.

Sprott (eds.) *Hormonal Control of Behaviour. Vol1.* Plenum Press, New York, pp. 205-72

_____ 1978. Neurobehavioural endocrine regulation in small mammal populations. In D.P. Snyder (ed.) *Populations of Small Mammals Under Natural Conditions.* University of Pittsburgh Press, Pittsburgh, pp. 143-58

_____ 1980. Endocrine factors in population regulation. In M.N. Cohen, R.S. Malpass and H.G. Klein (eds.) *Biosocial Mechanisms in Population Regulation.* Yale University Press, New Haven, pp. 55-115

_____ and Davis, D.E. 1964. Endocrines, behavior and population. *Science 146* : 1550-60

Clark, A.B. 1978. Sex ratio and local resource competition in a prosimian primate. *Science 201* : 163-5

Clark, M.M., Spencer, C.A. and Galef, B.G. 1986. Reproductive life history correlates of early and late sexual maturation in female Mongolian gerbils (*Meriones unguiculatus*). *Anim. Behav. 34* : 551-60

Clarke, J.R. 1981. Physiological problems of seasonal breeding in eutherian mammals. *Oxford Rev. Reprod. Biol. 3* : 244-312

_____ and Clulow, F.V. 1973. The effect of successive matings upon bank vole (*Clethrionomys glareolus*) and vole (*Microtus agrestis*). In H. Peters (ed.) *The Development and Maturation of the Ovary and its Functions.* Experta Medica, Amsterdam, pp. 160-70

Clulow, F.V. and Clarke, J.R. 1968. Pregnancy block in *Microtus agrestis* an induced ovulator. *Nature, Lond. 219* : 511

_____ and Langford, P.E. 1971. Pregnancy block in the meadow vole, *Microtus pennsylvanicus. J. Reprod. Fertil. 24* : 275-7

_____ , Franchetto, E.A. and Langford, P.E. 1982. Pregnancy failure in the red-backed vole, *Clethrionomys gapperi. J. Mamm. 63* : 499-500

Clutton-Brock, T.H. 1983. Selection in relation to sex. In D.S. Bendall (ed.) *Evolution from Molecules to Men.* Cambridge University Press, Cambridge, pp. 457-81

_____ 1984. Reproductive effort and terminal investment in iteroparous animals. *Amer. Natur. 123* : 212-29

_____ 1985. Birth sex ratios and the reproductive success of sons and daughters. In P.J. Greenwood, P.H. Harvey and M. Slatkin (eds.) *Evolution : Essays in Honour of John Maynard Smith.* Cambridge University Press, Cambridge, pp. 221-35

_____ and Albon, S.D. 1982. Parental investment in male and female offspring in mammals. In Kings' College Sociobiology Group (eds.) *Current Problems in Sociobiology.* Cambridge University Press, Cambridge, pp. 223-47

_____ , _____ and Guinness, F.E. 1981. Parental investment in male and female offspring in mammals. *Nature 289* : 487-9

_____ , _____ and _____ 1985. Parental investment and sex differences in juvenile mortality in birds and mammals. *Nature 313* : 131-3

_____ , _____ and _____ 1986. Great expectations: dominance, breeding success and offspring sex ratios in red deer. *Anim. Behav. 34* : 460-71

_____ , Guinness, F.E. and Albon, S.D. 1982. *Red Deer - Behavior and Ecology of Two Sexes.* Edinburgh University Press, Edinburgh

_____ , Harvey, P.H. and Rudder, B. 1977. Sexual dimorphism, socionomic sex ratio and body weight in primates. *Nature 269* : 797-800

Cockburn, A. 1978. The distribution of *Pseudomys shortridgei* (Muridae:Rodentia) and its relevance to that of other heathland Pseudomys. *Aust. Wildl. Res. 5* : 213-9

_____ 1981. Population regulation and dispersion of the smoky mouse, *Pseudomys fumeus.* II. Spring decline, breeding success and habitat heterogeneity. *Aust. J. Ecol. 6* : 255-67

_____ 1984. Environmental correlates of home ranges in cryptic small mammal species. *Acta Zool. Fennica 171* : 251-54

_____ 1985. Does dispersal increase as populations expand? *Oikos 44* : 367-8

_____, Braithwaite, R.W. and Lee, A.K. 1981. The response of the heath rat, *Pseudomys shortridgei*, to pyric succession: a temporally dynamic life history strategy. *J.Anim. Ecol. 50* : 649-66

_____, Lee, A.K. and Martin, R.W. 1983. Macrogeographic variation in litter size in *Antechinus* (Marsupialia: Dasyuridae). *Evolution 37* : 86-95

_____ and Lidicker, W.Z. 1983. Microhabitat heterogeneity and population ecology of an herbivorous rodent, *Microtus californicus. Oecologia (Berl). 59* : 167-77

_____, Scott, M.P. and Dickman, C.R. 1985. Sex ratio and intrasexual kin competition in mammals. *Oecologia (Berl). 66* : 427-9

_____, _____ and Scotts, D.J. 1985. Inbreeding avoidance and sex-biased dispersal in *Antechinus* spp. (Marsupialia, Dasyuridae). *Anim. Behav. 33* : 908-15

Cody, M.L. 1966. A general theory of clutch-size. *Evolution 20* : 174-84

Cole, F.R. and Batzli, G.O. 1978. Influence of supplemental feeding on a vole population. *J. Mamm. 59* : 809-19

_____ and _____ 1979. Nutrition and population dynamics of the prairie vole, *Microtus ochrogaster*, in central Illinois. *J. Anim. Ecol. 48* : 455-70

Cole, L.C. 1954. The population consequences of life history phenomena. *Q. Rev. Biol. 29* : 103-27

Colwell, R. 1981. Group selection is implicated in the evolution of female-biased sex ratios. *Nature 290* : 401-4

Comins, H.N. 1982. Evolutionarily stable strategies for localized dispersal in two dimensions. *J. theor. Biol. 94* : 579-606

_____, Hamilton, W.D. and May, R.M. 1980. Evolutionarily stable dispersal strategies. *J. theor. Biol. 82* : 205-30

Cooper, W.S. 1984. Expected time to extinction and the concept of fundamental fitness. *J. theor. Biol. 107* : 603-29

Coopersmith, C.B. and Banks, E.M. 1983. Effects of olfactory cues on sexual behaviour in the brown lemming, *Lemmus trimucronatus. J. comp. Psychol. 97* : 120-6

Coppola, D.M. and Vandenbergh, J.G. 1985. Effect of density, duration of grouping and age of urine stimulus on the puberty delay pheromone in female mice. *J. Reprod. Fert. 73* : 517-22

Crook, J.H. 1964. The evolution of social organization and visual communication in the weaver birds (Ploceinae). *Behaviour, Suppl. 10* : 1-178

_____ 1965. The adaptive significance of avian social organisations. *Symp. Zool. Soc. Lond. 14* : 181-218

Daketse, M. and Martinet, L. 1977. Effect of temperature on the growth and fertility of the field-vole, *Microtus arvalis*, raised in different day length and feeding conditions. *Ann. Biol. Anim. Biochem. Biophys. 17* : 713-21

Dapson, R.W. 1979. Phenologic influences on cohort-specific reproductive strategies in mice (*Peromyscus polionotus*). *Ecology 60* : 1125-31

_____, Feldman, A.T. and Pane, G. 1980. Differential rates of aging in natural populations of old-field mice (*Peromyscus polionotus*). *J. Geront. 35* : 39-44

Darwin, C. 1871. *The Descent of Man and Selection in Relation to Sex*. Murray, London

Davies, N.B. 1985. Cooperation and conflict among dunnocks, *Prunella modularis*, in a variable mating system. *Anim. Behav. 33* : 628-48

_____ and Houston, A.I. 1981. Owners and satellites : the economics of territory defence in the pied wagtail, *Motacilla alba. J. Anim. Ecol. 50* : 157-80

_____ and _____ 1983. Time allocation between territories and flocks and owner-satellite conflict in foraging pied wagtails, *Motacilla alba. J. Anim. Ecol. 52* : 621-34

_____ and _____ 1984. In J.R. Krebs and N.B. Davies (eds.) *Behavioural Ecology: An Evolutionary Approach* 2nd ed. Blackwell Scientific Publications, Oxford, pp. 148-69

_____ and Lundberg, A. 1984. Food distribution and a variable mating system in the dunnock, *Prunella modularis. J. Anim. Ecol.* 53 : 895-912

_____ and _____ 1985. The influence of food on time budgets and timing of breeding of the dunnock *Prunella modularis. Ibis 127* : 100-10

Davis, D.H.S. 1964. Ecology of wild rodent plagues. In D.H.S. Davis (ed.) *Ecological Studies in Southern Africa*. Dr. W. Junk, The Hague, pp. 301-14

Davis, H.N., Gray, G.D., Zerylnick, M. and Dewsbury, D.A. 1974. Ovulation and implantation in montane voles (*Microtus montanus*) as a function of varying amounts of copulatory stimulation. *Horm. Behav.* 5 : 383-8

Dawkins, R. 1976. *The Selfish Gene*. Oxford University Press, Oxford

_____ 1982. *The Extended Phenotype*. Freeman, Oxford.

_____ and Krebs, J.R. 1978. Animal signals : information or manipulation. In J.R. Krebs and N.B. Davies (eds.) *Behavioural Ecology : An Evolutionary Approach*, Blackwell Scientific Publications, Oxford, pp. 282-309

Dekker, H. 1975. A simple mathematical model of rodent population cycles. *J. Math. Biol.* 2 : 57-67

Demment, M.W. 1983. Feeding ecology and the evolution of body size of baboons. *Afr. J. Ecol. 21* : 219-33

_____ and Van Soest, P.J. 1985. A nutritional explanation for body-size patterns of ruminant and nonruminant herbivores. *Amer. Natur. 125* : 641-72

De Poorter, M. 1984. An experimental test of predictions from different hypotheses of self regulation in the snowshoe hare (*Lepus americanus* Erxleben,1977). Ph.D thesis, Faculteit der Wetenschappen, Vrije Universiteit, Brussels

Desy, E.A. and Thompson, C.F. 1983. Effects of supplemental food on a *Microtus pennsylvanicus* population in central Illinois. *J. Anim. Ecol.* 52 : 127-40

Dewsbury, D.A. 1982a. Pregnancy blockage following multiple-male copulation or exposure at the time of mating in deer mice, *Peromyscus maniculatus. Behav. Ecol. Sociobiol. 11* : 37-42

_____ 1982b. Ejaculate cost and male choice. *Amer. Natur. 119* : 601-10

_____ 1985a. Paternal behaviour in rodents. *Amer. Zool.* 25 : 841-52

_____ 1985b. Aggression, copulation, and differential reproduction of deer mice (*Peromyscus maniculatus*) in a semi-natural enclosure. *Behaviour 91* : 1-23

_____ and Hartung, T.G. 1982. Copulatory behavior of three species of *Microtus. J. Mamm. 63* : 306-9

_____ and Sawrey, D.K. 1984. Male capacity as related to sperm production, pregnancy initiation, and sperm competition in deer mice (*Peromyscus maniculatus*). *Behav. Ecol. Sociobiol. 16* : 37-47

Diamond, J.M. 1982. Big-bang reproduction and ageing in male marsupial mice. *Nature* 298 : 115-16

Dingle, H. and Hegmann, J.P.(eds.) 1982. *Evolution and Genetics of Life Histories*. Springer-Verlag, New York

Dobrowolska, A. and Gromadzka-Ostrowska, J. 1983. Influence of photoperiod on morphological parameters, androgen concentration, haematological indices and serum protein fractions in common vole (*Microtus arvalis,* Pall.). *Comp. Biochem. Physiol.* 74A : 427-33

Dobson, F.S. 1981. An experimental examination of an artificial dispersal technique. *J. Mamm.* 62 : 74-81

_____ 1982. Competition for mates and predominant juvenile male dispersal in mammals. *Anim. Behav. 30* : 1183-92

_____ 1985. The use of phylogeny in behavior and ecology. *Evolution 39* : 1384-8
_____ and Jones, W.T. 1985. Multiple causes of dispersal. *Amer. Natur. 126* : 855-8
Doty, R.L. 1974. A cry for the liberation of the female rodent : courtship and copulation in the Rodentia. *Psychol. Bull. 81* : 159-72
Drickamer, L.C. 1976. Effect of size and sex ratio of litter on the sexual maturation of female mice. *J. Reprod. Fertil. 46* : 369-74
_____ 1979. Acceleration and delay of first vaginal estrus in wild stocks of Mus musculus. *J. Mammal. 60* : 215-6
_____ 1981a. Selection for age of sexual maturation in mice and the consequences for population regulation. *Behav. Neur. Biol. 31* : 82-9
_____ 1981b. Acceleration and delay of first vaginal estrus in mice previously selcted for early and late sexual maturation. *J. Reprod. Fertil. 63* : 325-9
_____ 1982. Delay and acceleration of sexual maturation in female mice by urinary cues from other females. *Devl. Psychobiol. 15* : 433-42
_____ 1983a. Male acceleration of puberty in female mice (*Mus musculus*). *J. Comp. Pyschol. 97* : 191-200
_____ 1983b. Effect of period of grouping of donors and duration of stimulus exposure on delay of puberty in female mice by a urinary chemosignal from grouped females. *J. Reprod. Fertil. 69* : 723-7
_____ 1983c. Mice selected for age of puberty : reverse selection and reproductive characteristics. *Biol. Reprod. 28* : 843-48
_____ 1984a. Urinary chemosignals from mice (*Mus musculus*): acceleration and delay of puberty in related and unrelated young females. *J. Comp. Psychol. 89* : 414-20
_____ 1984b. Urinary chemosignals and puberty in female house mice: effects of photoperiod and food deprivation. *Psychol. Behav. 33* : 907-11
_____ 1984c. Effects of very small doses of urine on acceleration and delay of sexual maturation in female house mice. *J. Reprod. Fertil. 71* : 475-7
_____ 1984d. Seasonal variation in acceleration and delay of sexual maturation in female -mice by urinary chemosignals. *J. Reprod. Fertil. 72* : 55-5
Dueser, R.D., Wilson, M.L. and Rose, R.K. 1981. Attributes of dispersing meadow voles in open-grid populations. *Acta Theriol. 26* : 139-69
Dunbar, R. 1984. The ecology of monogamy. *New Scientist 30 August 1984* : 12-15
Egid, K. and Lenington, S. 1985. Responses of male mice to odors of females: effects of T- and H-2 -locus genotype. *Behav. Genet. 15* : 287-95
Ekman, J. 1984a. Density-dependent seasonal mortality and population fluctuations of the temperate-zone willow tit (*Parus montanus*). *J. Anim. Ecol. 53* : 119-34
_____ 1984b. Stability and persistence of an age-structured avian population in a seasonal environment. *J. Anim. Ecol. 53* : 135-46
Ellstrand, N.C. 1984. Multiple paternity within the fruits of the wild radish, *Raphanus sativus. Amer. Natur. 123* : 819-28
Elton, C.S. 1942. *Voles, Mice and Lemmings: Problems of Population Dynamics.* Oxford University Press, London
_____ and Nicholson, M. 1942. The ten-year cycle in numbers of the lynx in Canada. *J. Anim. Ecol. 11* : 215-44
Elwood, R.W. 1986. What makes male mice paternal? *Behav. Neur. Biol. 46* : 54-63
_____ and Ostermeyer, M.C. 1984. Does copulation inhibit infanticide in male rodents? *Anim. Behav. 32* : 293-4
Emlen, J.M. 1984. *Population Biology : The Evolution of Population Dynamics and Behavior.* Macmillan Publishing Company, New York
Emlen, S.T. and Oring, L.W. 1977. Ecology, sexual selection, and the evolution of mating systems. *Science 197* : 215-23
Endler, J.A. 1977. *Geographic Variation, Speciation, and Clines.* Princeton University

Press, Princeton

Eriksson, M. 1984. Winter breeding in three rodent species, the bank vole *Clethrionomys glareolus*, the yellow-necked mouse *Apodemus flavicollis* and the wood mouse *Apodemus sylvaticus* in southern Sweden. *Holarctic Ecology 7* : 428-9

Erlinge, S., Göransson, G., Högstedt, G., Liberg, O., Nilsson, I.N., Nilsson, T., von Schantz, T and Sylvén, M. 1983. Predation as a regulating factor in small rodent populations in southern Sweden. *Oikos 40* : 36-52

_____ , _____ , _____ , Jansson, G., Liberg, O., Loman, J., Nilsson, I.N., von Schantz, T and Sylvén, M. 1984. Can vertebrate predators regulate their prey? *Amer. Natur. 123* : 125-33

Evans, D.M. 1973. Seasonal variations in the body composition and nutrition of the vole *Microtus agrestis*. *J. Anim. Ecol. 42* : 1-18

Facemire, C.F. and Batzli, G.O. 1983. Suppression of growth and reproduction by social factors in microtine rodents : tests of two hypotheses. *J. Mamm. 64* : 152-6

Fahrig, L. and Merriam, G. 1985. Habitat patch connectivity and population survival. *Ecology 66* : 1762-8

Fairbairn, D.J. 1977. The spring decline in deer mice : death or dispersal? *Can. J. Zool. 55* : 84-92

Falconer, D.S. 1981. *Introduction to Quantitative Genetics*. 2nd ed. Longman, London and New York

Fedyk, A. and Gebcynski, M. 1980. Genetic changes in seasonal generations of the bank vole. *Acta Theriol. 25* : 475-85

Feeny, P. 1975. Biochemical coevolution between plants and their herbivores. In L.E. Gilbert and P.H. Raven (eds.) *Coevolution of Animals and Plants*. University of Texas Press, Austin, pp. 3-19

_____ 1976. Plant apparency and chemical defense. *Recent Adv. Phytochem. 10* : 1-40

Felsenstein, J. 1985. Phylogenies and the comparative method. *Amer. Natur. 125* : 1-15

Fenyuk, B.K. 1937. The influence of agriculture on the numbers of mouse-like rodents and the biological foundations of their control [in Russian]. *Vest. Mikrob. Epidem. Parazit. 16* : 478-92

Ferns, P.N. 1979. Growth, reproduction and residency in a declining population of *Microtus agrestis*. *J. Anim. Ecol. 48* : 739-58

Fetcher, N., Beatty, T.F., Mullinax, B. and Winkler, D.S. 1984. Changes in arctic tussock tundra thirteen years after fire. *Ecology 65* : 1332-3

Findlay, C.S. and Cooke, F. 1983. Genetic and environmental components of clutch size variance in a wild population of lesser snow geese (*Anser caerulescens caerulescens*). *Evolution 37* : 724-34

Finerty, J.P. 1980. *The Population Ecology of Cycles in Small Mammals: Mathematical Theory and Biological Fact*. Yale University Press, New Haven and London

Fisher, R.A. 1930. *The Genetical Theory of Natural Selection*. Clarendon Press, Oxford

Fitzgerald, B.M. 1977. Weasel predation on a cyclic population of the montane vole (*Microtus montanus*) in California. *J. Anim. Ecol. 46* : 367-97

FitzGerald, R.W. and Madison, D.M. 1983. Social organization of a free-ranging population of pine voles, *Microtus pinetorum*. *Behav. Ecol. Sociobiol 13* : 183-7

Fleischer, R.C. and Johnston, R.F. 1984. The relationships between winter climate and selection on body size of house sparrows. *Can. J. Zool. 62* : 405-10

Flowerdew, J.R. 1972. The effects of supplementary food on a population of wood mice (*Apodemus sylvaticus*). *J. Anim. Ecol. 41* : 553-66

Flux, J.E.C. and Flux, M.M. 1982. Artificial selection and gene flow in wild starlings (*Sturnus vulgaris*). *Naturwissenschaften 69* : 96-7

Foltz, D.W. 1981. Genetic evidence for long-term monogamy in a small rodent, *Peromyscus polionotus*. *Amer. Natur. 117* : 665-75

_____ and Hoogland, J.L. 1981. Analysis of the mating system in the black-tailed prairie dog (*Cynomys ludovicianus*) by likelihood of paternity. *J. Mamm.* 62 : 706-12

Ford, R.G. and Krumme, D.W. 1979. The analysis of space use patterns. *J. theor. Biol.* 76 : 125-55

_____ and Pitelka, F.A. 1984. Resource limitation in populations of the California vole. *Ecology* 65 : 122-36

Fox, J.F. 1978. Forest fires and the snowshoe hare-Canada lynx cycle. *Oecologia (Berl.)* 31 : 329-74

_____ and Bryant, J.P. 1984. Instability of the snowshoe hare and woody plant interaction. *Oecologia (Berl.)* 63 : 128-35

Fox, L.R. 1981. Defense and dynamics in plant-herbivore systems. *Amer. Zool.* 21 : 853-64

Frank, F. 1954. Beitrage zur Biologie der Feldmaus *Microtus arvalis* I. Gehegeversuche. *Zool. Jahrb 3 , 82* : 354-404

Frankiewicz, J. and Marchlewska-Koj, A. 1985. Effects of conspecifics on sexual maturation in female European pine voles (*Pitymys subterraneus*). *J. Reprod. Fertil.* 74 : 153-6

Fredga, K.A., Gropp, A., Winking, H. and Frank, F. 1976. Fertile XX- and XY- type females in the wood lemming (*Myopus schisticolor*). *Nature* 261 : 225-7

_____ , _____ , _____ and _____ 1977. A hypothesis explaining the exceptional sex ratio in the wood lemming (*Myopus schisticolor*). *Hereditas* 85 : 101-4

Freeland, W.J. 1974. Vole cycles : another hypothesis. *Amer. Natur.* 108 : 238-45

Fretwell, S.D. 1972. *Populations in a Seasonal Environment.* Princeton University Press, Princeton, New Jersey

_____ and Lucas, H.L. 1970. On territorial behavior and other factors influencing habitat distribution in birds. I. Theoretical development. *Acta Biotheor.* 19 : 16-36

Fritz, R.S. 1979. Consequences of insular population sructure : distribution and extinction of spruce grouse populations. *Oecologia (Berl.)* 42 : 57-65

Frogner, K.J. 1980. Variable developmental period : Intraspecific competition models with conditional age-specific maturity and mortality schedules. *Ecology* 61 : 1099-1106

Fuentes, S. and Dewsbury, D.A. 1984. Copulatory behavior in voles (*Microtus montanus* and *M. ochrogaster*) in multiple-female test situations. *J. comp. Physiol. Psychol.* 98 : 45-53

Gadgil, M. 1971. Dispersal: population consequences and evolution. *Ecology* 52 : 253-61

_____ and Bossert, W. 1970. Life history consequences of natural selection. *Amer. Natur.* 104 : 1-24

Gaines, M.S. 1985. Genetics. In R.H. Tamarin (ed.), *Biology of New World Microtus.* Special Publication No. 8 , American Society of Mammalogists, pp. 845-883

_____ , Baker, C.L. and Vivas, A.M. 1979. Demographic attributes of dispersing southern bog lemmings (*Synaptomys cooperi*) in eastern Kansas. *Oecologia (Berl.)* 40 : 91-101

_____ and Gorman, W.L. 1985. Are transferrin and leucine aminopeptidase electromorphs reliable genetic markers in the prairie vole, *Microtus ochrogaster*? *Oecologia (Berl.)* 66 : 74-86

_____ and Johnson, M.L. 1982. Home range size and population dynamics in the prairie vole, *Microtus ochrogaster*. *Oikos* 39 : 63-70

_____ and _____ 1984. A multivariate study of the relationship between dispersal and demography in populations of *Microtus ochrogaster* in eastern Kansas. *Amer. Midl. Natur.* 111 : 223-33

_____ and McClenaghan, L.R. 1980. Dispersal in small mammals. *Ann. Rev. Ecol.*

Syst. 11 : 163-96
———, ——— and Rose, R. 1978. Temporal patterns of allozymic variation in fluctuating populations of *Microtus ochrogaster*. *Evolution 32* : 723-9
———, Rose, R.K. and McClenaghan, L.R. 1978. The demography of *Synaptomys cooperi* populations in eastern Kansas. *Can. J. Zool. 55* : 1584-94
———, Schaffer, W.M. and Rose, R.K. 1979. Additional comments on reproductive strategies and population fluctuations in microtine rodents. *Ecology 60* : 1284-6
——— and Whittam, T.S. 1980. Genetic changes in fluctuating vole populations: selective vs nonselective forces. *Genetics 96* : 767-78
———, Vivas, A.M. and Baker, C.L. 1979. An experimental analysis of dispersal in fluctuating vole populations : demographic parameters. *Ecology 60* : 814-29
Garsd, A. and Howard, W.E. 1981. A 19-year study of microtine population fluctuations using time-series analysis. *Ecology 62* : 930-7
Garten, C.T. 1976. Relationships between aggressive behavior and genetic heterozygosity in the old field mouse, *Peromyscus polionotus*. *Evolution 30* : 59-73
Gashwiler, J.S. 1972. Life history notes on the Oregon vole, *Microtus oregoni. J. Mamm. 53* : 558-69
Gavish, L., Carter, C.S. and Getz, L.L. 1981. Further evidences for monogamy in the prairie vole. *Anim. Behav. 29* : 955-7
———, ——— and ——— 1983. Male-female interactions in prairie voles. *Anim. Behav. 31* : 511-17
———, Hofmann, J.E. and Getz, L.L. 1984. Sibling recognition in the prairie vole, *Microtus ochrogaster*. *Anim. Behav. 32* :362-6
Getty, T. 1981. Competitive collusion: the preemption of competition during the sequential establishment of territories. *Amer. Natur. 118* : 426-31
Getz, L.L. 1978. Speculation on social structure and population cycles of microtine cycles. *Biologist (Phi Sigma Soc.) 60* : 134-47
——— 1985. Habitats. In R.H. Tamarin (ed.), *Biology of New World* Microtus. Special Publication No. 8, American Society of Mammalogists, pp. 286-309
——— and Carter, C.S. 1980. Social organization in *Microtus ochrogaster* populations. *Biologist 62* : 56-69
———, ——— and Gavish, L.L. 1981. The mating system of the prairie vole, *Microtus ochrogaster* : field and laboratory evidence for pair-bonding. *Behav. Ecol. Sociobiol. 8* : 189-94
——— and Hofmann, J.E. 1986. Social organization in free-living prairie voles, *Microtus ochrogaster*. *Behav. Ecol. Sociobiol. 18* : 241-9
Gibson, J.B. and Oakeshott, J.G. 1980. *Genetic Studies of* Drosophila *Populations*. Australian National University Press, Canberra
Giesel, J.T. 1976. Reproductive strategies as adaptations to life in temporally heterogeneous environments. *Ann. Rev. Ecol. Syst. 7* : 57-79
———, Murphy, P.A. and Manlove, M.N. 1982. The influence of temperature on genetic interrelationships of life history traits in a population of *Drosophila melanogaster* : what tangled data sets we weave. *Amer. Natur. 119* : 464-79
Gilbert, B.S. and Krebs, C.J. 1981. Effects of extra food on *Peromyscus* and *Clethrionomys* populations in the southern Yukon. *Oecologia (Berl.) 51* : 326-31
——— and ——— 1984. Competition between *Peromyscus maniculatus* and other small rodents in boreal forest of the southern Yukon territory. *Acta Zool. Fennica 172* : 51-6
———, ———, Talarico, D. and Cichowski, D.B. 1986. Do *Clethrionomys rutilus* females suppress maturation of juvenile females? *J. Anim. Ecol. 55* : 543-52
Gileva, E.A. 1980. Chromosomal diversity and an aberrant system of sex determination in the Arctic lemming, *Dicrostonyx torquatus* Pallas (1779). *Genetica 52/53* : 99-103

_____ and Chebotar, N.A. 1979. Fertile XO males and females in the varying lemming, *Dicrostonyx torquatus* Pall. (1779). *Heredity 42* : 67-77

_____, Benenson, I.E., Konopistseva, L.A., Puchkov, V.F. and Makarenets, I.A. 1982. XO females in the varying lemming, *Dicrostonys torquatus* : reproductive performance and its evolutionary significance. *Evolution 36* : 601-9

Gill, A.E. 1976. White spotting in the California vole. *Heredity 37* : 113-28

_____ 1977. Maintenance of polymorphisms in an island population of the California vole, *Microtus californicus*. *Evolution 31* : 512-25

Gill, D.E. 1974. Intrinsic rates of increase, saturation densities, and competitive ability. II. The evolution of competitive ability. *Amer. Natur. 108* : 103-16

_____ 1978a. On selection at high population density. *Ecology 59* : 1289-91

_____ 1978b. The metapopulation dynamics of the red-spotted newt, *Notophthalmus viridescens* (Rafinesque). *Ecol. Monogr. 48* : 145-66

Gillespie, J.H. 1974. Natural selection for within-generation variance in offspring number. *Genetics 76* : 601-6

_____ 1975. Natural selection for within-generation variance in offspring number, II. Discrete haploid models. *Genetics 81* : 403-13

_____ 1977. Natural selection for variances in offspring numbers : a new evolutionary principle. *Amer. Natur. 111* : 1010-14

Gilmore, R.M. 1947. Cycle behavior and economic importance of the rata-maca (*Oryzomys*) in Peru. *J. Mamm. 28* : 231-41

Gipps, J.H.W. 1985. Spacing behaviour and male reproductive ecology in voles of the genus Clethrionomys. *Ann. Zool. Fennici 22* : 343-51

_____, Flynn, M.P., Gurnell, J. and Healing, T.D. 1985. The spring decline in populations of the bank vole, *Clethrionomys glareolus*, and the role of female density. *J. Anim. Ecol. 54* : 351-8

_____ and Jewell, P.A. 1979. Maintaining populations of bank voles, *Clethrionomys glareolus*, in large outdoor enclosures, and measuring the response of population variables to the castration of voles. *J. Anim. Ecol. 48* : 535-55

Glass, G.E. 1986. Models of infanticide: a reply to Hausfater. *Anim. Behav. 34* : 619-21

_____, Holt, R.D. and Slade, N.A. 1985. Infanticide as an evolutionarily stable strategy. *Anim. Behav. 33* : 384-91

Glazier, D.S. 1985. Energetics of litter size in five species of Peromyscus with generalizations for other mammals. *J. Mamm. 66* : 629-42

Gleason, P.E., Michael, S.D. and Christian, J.J. 1980. Aggressive behavior during the reproductive cycle of female *Peromyscus leucopus*: effects of encounter site. *Behav. Neur. Biol. 29* : 506-11

Gliwicz, J. 1980. Island populations of rodents: their organization and functioning. *Biol. Rev. 55* : 109-38

_____ 1987. Patterns of dispersal in stable populations of small rodents. In N.C. Stenseth and W.Z. Lidicker (eds.) *Animal Dispersal : Small Mammals as a Model*. Chapman and Hall, London, pp. 000-000

Goldberg, M., Tabroff, N.R. and Tamarin, R.H. 1980. Nutrient variation in relation to beach vole feeding. *Ecology 61* : 1029-33

Goodman, D. 1979. Regulating reproductive effort in a changing environment. *Amer. Natur. 113* : 735-48

_____ 1984. Risk spreading as an adaptive strategy in iteroparous life histories. *Theor. Popul. Biol. 25* : 1-20

Gosling, L.M., Baker, S.J. and Wright, K.M.H. 1984. Differential investment by female coypus (*Myocastor coypus*) during lactation. *Symp. Zool. Soc. Lond. 51* : 273-300

Gould, S.J. and Lewontin, R.C. 1979. The spandrels of San Marco and the Panglossian

paradigm: a critique of the adaptionist programme. *Proc. R. Soc. Lond.*
B205 : 581-98
Graf, R.P. 1985. Social organization of snowshoe hares. *Can. J. Zool. 63* : 468-74
Grafen, A. 1984. Natural selection, kin selection and group selection. In J.R. Krebs and N.B. Davies (eds.) *Behavioural Ecology: An Evolutionary Approach* 2nd edn. Blackwell Scientific Publications, Oxford, pp. 62-84
Grange, W.B. 1932. Observations of the snowshoe hare, *Lepus americanus* Phaenoutus Allen. *J. Mamm 13* : 1-19
Grant, B.R. 1984. The significance of song variation in a population of Darwin's finches. *Behaviour 89* : 90-116
Grant, P. R. 1975. Population performance of *Microtus pennsylvanicus* confined to woodland habitat, and a model of habitat occupancy. *Can. J. Zool. 53* : 1447-65
Gray, G.D., Davis, H.N., Kenney, A.M. and Dewsbury, D.A. 1976. Effect of mating on plasma levels of LH and progesterone in montane voles (*Microtus montanus*). *J. Reprod. Fertil. 47* : 89-91
_____ , _____ , Zerylnick, M. and Dewsbury, D.A. 1974. Oestrus and induced ovulation in montane voles. *J. Reprod. Fertil. 38* : 193-6
_____ and Dewsbury, D.A. 1973. A quantitative description of copulatory behaviour in prairie voles (*Microtus ochrogaster*). *Brain Behav. Evol. 8* : 837-52
Gray, G.D. and Dewsbury, D.A. 1975. A quantitative description of the copulatory behaviour of meadow voles (*Microtus pennsylvanicus*). *Anim. Behav. 23* : 261-7
Greenberg, L. 1979. Genetic component of bee odor in kin recognition. *Science 206* : 1095-7
Greenslade, P.J.M. 1983. Adversity selection and the habitat templet. *Amer. Natur. 122* : 352-65
Greenwood, P.J. 1980. Mating systems, philopatry and dispersal in birds and mammals. *Anim. Behav. 28* : 1140-62
_____ 1983. Mating systems and the evolutionary consequences of dispersal. In I.R. Swingland and P.J. Greenwood (eds.) *The Ecology of Animal Movement.* Clarendon Press, Oxford, pp. 116-31
_____ and Harvey, P.H. 1982. The natal and breeding dispersal of birds. *Ann. Rev. Ecol. Syst. 13* : 1-21
_____ , _____ and Perrins, C.M. 1979. The role of dispersal in the great tit (*Parus major*): the causes, consequences and heritability of natal dispersal. *J. Anim. Ecol. 48* : 123-42
Grime, J.P. 1977. Evidence for the existence of three primary strategies in plants and its relevance to ecological and evolutionary theory. *Amer. Natur. 111* : 1169-94
_____ 1979. *Plant Reproductive Strategies and Vegetation Processes.* Wiley, Chichester
Grocock, C.A. 1979. Testis development in the vole, *Microtus agrestis*, subjected to long or short photoperiods from birth. *J. Reprod. Fertil. 55* : 423-7
_____ 1981. Effects of different photoperiods on testicular weight changes in the vole, *Microtus agrestis. J. Reprod. Fertil. 62* : 25-32
Grodzinski, W., Makomaska, M., Tertil, R. and Weiner, J. 1977. Bioenergetics and total impact of vole populations. *Oikos 29* : 494-510
Gropp, A., Winking, H., Frank, F., Noack, G. and Fredga, K. 1976. Sex-chromosome aberrations in wood lemmings (*Myopus schisticolor*). *Cytogenet. Cell Genet. 17* : 343-58
Gross, J.E., Wang, Z. and Wunder, B.A. 1985. Effects of food quality and energy needs: changes in gut morphology and capacity of *Microtus ochrogaster. J. Mamm. 66* : 661-7
Gruder-Adams, S. and Getz, L.L. 1985. Comparison of the mating system and paternal behavior in *Microtus ochrogaster* and *M. pennsylvanicus. J. Mamm. 66* : 165-7

Gulyeava, I.P. and Olenev, G.V. 1980. Variations in the electrophoretic pattern of serum transferrins in the common red-backed vole as a function of the animals physiological state. *Soviet Journal of Ecology 10* : 535-9

Gurnell, J. 1978. Seasonal changes in numbers and male behavioural interaction in a population of wood mice, *Apodemus sylvaticus*. *J. Anim. Ecol. 47* : 741-55

Gurney, W.S.C. and Nisbet, R.M. 1979. Ecological stability and social hierarchy. *Theor. Popul. Biol. 16* : 48-80

Gustafsson, T.O. 1985. Sexual maturation in *Clethrionomys*. *Ann. Zool. Fennici* 22 : 303-8

―――― and Andersson, C.B. 1980. Social environment and sexual maturation in male bank voles, *Clethrionomys glareolus* (Schreber, 1780). *Säugetierk. Mitteil.* 28 : 310-2

――――, ―――― and Nyholm, N.E.I. 1983. Comparison of sensitivity to social suppression of sexual maturation in captive bank voles, *Clethrionomys glareolus*, originating from populations with different degrees of cyclicity. *Oikos 41* : 250-4

――――, ―――― and Westlin, L.M. 1983. Reproduction in laboratory colonies of bank vole, *Clethrionomys glareolus*, originating from populations with different degrees of cyclicity. *Oikos 40* : 182-8

―――― and Batzli, G.O. 1985. Effects of diet and origin of animals on growth and reproduction of *Clethrionomys glareolus*. *Ann. Zool. Fennici 22* : 273-6

Guthrie, R.D. 1971. Factors regulating the evolution of microtine tooth complexity. *Z. Saugetierk. 36* : 37-54

Gyllensten, U. 1985. Temporal allozyme frequency changes in density fluctuating populations of willow grouse (*Lagopus lagopus* L.). *Evolution 39* : 115-21

Gyug, L.W. and Millar, J.S. 1981. Growth of seasonal generations in three natural populations of *Peromyscus*. *Can. J. Zool. 59* : 510-14

Hairston, N.G. and Munns, W.R. 1984. The timing of copepod diapause as an evolutionary stable strategy. *Amer. Natur. 123* : 733-51

Halfpenny, J.C. and Ingraham, K.P. 1983. Growth and development of heather voles. *Growth 47* : 437-45

Hall, S.J.G. and Semeneoff, R. 1985. Plasma esterase polymorphism in the bank vole, *Clethrionomys glareolus*, in Britain. *J. Zool. Lond.(A) 207* : 213-22

Halliday, T.R. 1983. The study of mate choice. In P. Bateson (ed.) *Mate Choice*. Cambridge University Press, Cambridge, pp. 3-32

Hamilton, W.D. 1964. The genetical evolution of social behaviour. *J. Theor. Biol.* 7 : 1-52

―――― 1967. Extraordinary sex ratios. *Science 156* : 477-88

―――― 1970. Selfish and spiteful behaviour in an evolutionary model. *Nature* 228 : 1218-20

―――― 1982. Pathogens as causes of genetic diversity in their host organisms. In R.M. Anderson and R.M. May (eds.) *Population Biology of Infectious Diseases*. Springer-Verlag, Berlin, pp. 269-96

―――― and May, R.M. 1978. Dispersal in stable habitats. *Nature 269* : 578-81

―――― and Zuk, M. 1982. Heritable true fitness and bright birds : a role for parasites? *Science 218* : 384-7

Hamilton, W.J. 1941. Reproduction of the field mouse *Microtus pennsylvanicus* (Ord.). *Cornell Univ. Agric. Exp. Sta. Mem. 237* : 1-23

―――― 1962. Reproductive adaptations of the red tree mouse. *J. Mamm. 43* : 486-504

Hanken, J. and Sherman, P.W. 1981. Multiple paternity in Belding's ground squirrels. *Science 212* : 351-3

Hannon, S.J. 1983. Spacing and breeding density of willow ptarmigan in response to experimental alteration of the sex ratio. *J. Anim. Ecol. 52* : 807-20

_____ 1984. Factors limiting polygyny in the willow ptarmigan. *Anim. Behav.* 32 : 153-61

_____ and Roland, J. 1984. Morphology and territory acquisition in willow ptarmigan. *Can. J. Zool.* 62 : 1502-6

Hannon, S.J. and Smith, J.N.M. 1984. Factors influencing age-related reproductive success in the willow ptarmigan. *Auk 101* : 848-54

_____ , Sopuck, L.G. and Zwickel, F.C. 1982. Spring movements of female blue grouse: evidence for socially induced delayed breeding in yearlings. *Auk 99 :* 687-94

Hansson, L. 1971. Small rodent food, feeding and population dynamics. A comparison between granivorous and herbivorous species in Scandinavia. *Oikos* 22 : 183-98

_____ 1972. Seasonal changes in physiology and nutrition of herbivorous small rodents. *Aquilo: Ser. Zool. 13* : 53-5

_____ 1977. Spatial dynamics of field voles *Microtus agrestis* in heterogeneous landscapes. *Oikos 29* : 539-44

_____ 1979a. On the importance of habitat heterogeneity in northern regions for the breeding population densities of homeotherms : a general hypothesis. *Oikos 33 :* 182-9

_____ 1979b. Condition and diet in relation to habitat in bank voles *Clethrionomys glareolus*: population or community approach? *Oikos 33 :* 55-63

_____ 1979c. Sex ratio in small mammal populations as affected by the pattern of fluctuations. *Acta theriol. 23 :* 203-12

_____ 1982. Experiments on habitat selection in voles : implications for the inverse distribution of two common European species. *Oecologia (Berl.) 52 :* 246-52

_____ 1984a. Predation as a factor causing extended low densities in microtine cycles. *Oikos 43 :* 255-6

_____ 1984b. Winter reproduction of small mammals in relation to food conditions and population dynamics. In J.F. Merritt (ed.), *Winter Ecology of Small Mammals*, Special Publication No. 10 : Carnegie Museum of Natural History, Pittsburgh, pp. 225-34

_____ 1984c. Composition of cyclic and non-cyclic vole populations : on the causes of variation in individual quality among *Clethrionomys glareolus* in Sweden. *Oecologia (Berl.) 63* : 199-206

_____ 1985a. *Clethrionomys* food: generic, specific and regional characteristics. *Ann. Zool. Fennici 22* : 315-8

_____ 1985b. Geographic differences in bank voles *Clethrionomys glareolus* in relation to ecogeographical rules and possible demographic and nutritive strategies. *Ann. Zool. Fennici* 22 : 319-28

_____ 1987. Small mammal dispersal in pest management and conservation. In N.C. Stenseth and W.Z. Lidicker (eds.). *Animal Dispersal : Small Mammals as a Model*. Chapman and Hall, London, pp. 000-000

_____ and Hentonnen, H. 1985a. Gradients in density variations of small rodents: the importance of latitude and snow cover. *Oecologia (Berl.) 67* : 394-402.

_____ and _____ 1985b. Regional differences in cyclicity and reproduction in *Clethrionomys* species : are they related? *Ann. Zool. Fennici 22 :* 277-88

_____ and Larsson, T-B. 1978. Vole diet on experimentally managed reforestation areas in northern Sweden. *Holarctic Ecology 1 :* 16-26

Hare, S. and Robertson, O.H. 1959. Changes in plasma 17-hydroxycorticosteroids accompanying sexual maturation and spawning of the Pacific salmon (*Oncorhynchus tschawytscha*) and rainbow trout (*Salmo gairdnerii*). *Proc. natl. Acad. Sci. USA* 45 : 886-93

Harper, J.L. 1982. After description. In E.I. Newman (ed.) *The Plant Community as a Working Mechanism*. Blackwell Scientific Publications, Oxford, pp 11-25

Harris, M.A. and Murie, J.O. 1984. Inheritance of nest sites in female Columbian ground squirrels. *Behav. Ecol. Sociobiol. 15* : 97-102

Hart, A. and Begon, M. 1982. The status of general reproductive-strategy theory, illustrated in winkles. *Oecologia (Berl.) 52* : 37-42

Hartung, T.G. and Dewsbury, D.A. 1978. A comparative analysis of copulatory plugs in muroid rodents and their relationship to copulatory behaviour. *J. Mamm. 59* : 717-23

Harvey, P.H. 1985. Intrademic group selection and the sex ratio. In R.M. Sibly and R.H. Smith (eds.) *Behavioural Ecology: Ecological Consequences of Adaptive Behaviour.* Blackwell Scientific Publications, Oxford, pp. 59-71

_____ , Partridge, L. and Nunney, L. 1985. Group selection and the sex ratio. *Nature 313* : 10-11

_____ and Ralls, K. 1986. Do animals avoid incest? *Nature 320* : 575-6

Hasler, M.J. and Banks, E.M. 1975. The influence of mature males on sexual maturation in female collared lemmings (*Dicrostonyx groenlandicus*). *J. Reprod. Fertil. 42* : 583-6

_____ , Buhl, A.E. and Banks, E.M. 1976. The influence of photoperiod on growth and sexual function in male and female collared lemmings (*Dicrostonyx groenlandicus*). *J. Reprod. Fertil. 46* : 323-9

_____ and Nalbandov, A.B. 1974. The effect of weanling and adult males on sexual maturation in female voles (*Microtus ochrogaster*). *Gen. Comp. Endocrin. 23* : 237-8

Hassell, M.P. and May, R.M. 1985. From individual behaviour to population dynamics. In R.M. Sibly and R.H. Smith (eds.) *Behavioural Ecology: Ecological Consequences of Adaptive Behaviour.* Blackwell Scientific Publications, Oxford, pp. 3-32

Hastings, A. 1982. Dynamics of a single species in a spatially varying habitat: The stabilizing role of high dispersal rates. *J. Math. Biol. 16* : 49-55

_____ 1983a. Age-dependent predation is not a simple process. I. Continuous time models. *Theor. Popul. Biol. 23* : 347-62

_____ 1983b. Can spatial variation alone lead to selection for dispersal? *Theor. Popul. Biol. 24* : 244-51

_____ 1984. Evolution of a seasonal environment: simplicity lost? *Evolution 38* : 350-8

Haukioja, E. 1980. On the role of plant defences in the fluctuation of herbivore populations. *Oikos 35* : 201-13

_____ and Hakala, T. 1975. Herbivore cycles and periodic outbreaks. Formulation of a general hypothesis. *Rep. Kevo Subarctic Res. Stat. 12* : 1-9

Hausfater, G. 1984. Infanticide in langurs : Strategies, counterstrategies and parameter values. In G.Hausfater and S. Hrdy (eds.) *Infanticide : Comparative and Evolutionary Perspectives.* Aldine, New York, pp. 257-82

_____ 1986. Convergent models : evidence of a robust theory of infanticide. *Anim. Behav. 34* : 617-9

_____ and Hrdy, S.B. (eds.) 1984. *Infanticide : Comparative and Evolutionary Perspectives.* Aldine Publishing Company, New York

Heckel, D.G. and Roughgarden, J. 1980. A species near its equilibrium size in a fluctuating environment can evolve a lower intrinsic rate of increase. *Proc. Natl. Acad. Sci. USA 77* : 7497-500

Hentonnen, H. 1985. Predation causing extended low densities in microtine cycles: further evidence from shrew dynamics. *Oikos 45* : 156-7

_____ , McGuire, A.D. and Hansson, L. 1985. Comparisons of amplitudes and frequencies (spectral analyses) of density variation in long-term data sets of *Clethrionomys* species. *Ann. Zool. Fenn. 22* : 221-7

_____, Tast, J., Viitala, J. and Kaikasulo, A. 1984. Ecology of cyclic rodents in northern Finland. *Memoranda Soc. Fauna Flora Fennica* 60 : 84-82

Heske, E.J. and Nelson, R.J. 1984. Pregnancy interruption in *Microtus ochrogaster*: laboratory artifact or field phenomenon? *Biol. Reprod.* 31 : 97-103

_____, Ostfeld, R.S. and Lidicker, W.Z. 1984. Competitive interactions between *Microtus californicus* and *Reithrodontomys megalotis* during two peaks of *Microtus* abundance. *J. Mamm.* 65 : 271-80

Hestbeck, J. 1982. Population regulation of cyclic small mammals : the social fence hypothesis. *Oikos* 39 : 157-63

Hickie, J.P. Lavigne, D.M. and Woodward, W.D. 1982. Vitamin D and winter reproduction in the collared lemming, *Dicrostonyx groenlandicus*. *Oikos* 39 : 71-6

Hilborn, R. 1975. Similarities in dispersal tendency among siblings in four species of voles (*Microtus*). *Ecology* 56 : 1221-5

Hirshfield, M.F. and Tinkle, D.W. 1975. Natural selection and the evolution of reproductive effort. *Proc. Nat. Acad. Sci. USA* 72 : 2227-31

Hoffmeyer, I. 1982. Responses of female bank voles (*Clethrionomys glareolus*) to dominant vs subordinate conspecific males and to urine odors from dominant vs subordinate males. *Behav. Neur. Biol.* 36 : 178-88

Hofmann, J.E., Getz, L.L. and Klatt, B.J. 1982. Levels of male aggressiveness in fluctuating populations of *Microtus ochrogaster* and *M. pennsylvanicus*. *Can. J. Zool.* 60 : 898-912

Högstedt, G. 1980. Evolution of clutch size in birds: adaptive variation in relation to territory quality. *Science* 210 : 1148-50

Holekamp, K.E. 1984a. Dispersal in ground-dwelling sciurids. In J.O. Murie and G.R. Michener (eds.) *The Biology of Ground-Dwelling Squirrels*. University of Nebraska Press, Lincoln and London, pp. 297-320

_____ 1984b. Natal dispersal in Belding's ground squirrels (*Spermophilus beldingi*). *Behav. Ecol. Sociobiol.* 16 : 21-30

Holisová, V. 1959. Die Nahrung der Feldmaus. In J. Kratchovil (ed.) *Microtus arvalis*. Prague, pp. 100-129

_____ 1965. The food of the water vole, *Arvicola terrestris*, in the agrarian environment of south Moravia. *Zool. Listy* 14 : 209-18

_____ 1970. Trophic requirements of the water vole, *Arvicola terrestris* Linn., on the edge of stagnant waters. *Zool. Listy* 19 : 221-33

_____ 1975. The foods eaten by rodents in reed swamps of Nesyt fishpond. *Zool. Listy* 24 : 223-37

_____ 1976. The food eaten by the water vole (*Arvicola terrestris*) in gardens. *Zool. Listy* 25 : 193-208

Hölldobler, B. and Lumsden, C.J. 1980. Territorial strategies in ants. *Science* 210 : 732-9

Holmes, W.G. 1986. Identification of paternal half-siblings by captive Belding's ground squirrels. *Anim. Behav.* 34 : 321-7

_____ and Sherman, P.W. 1982. The ontogeny of kin recognition in two species of ground squirrels. *Amer. Zool.* 22 : 491-517

_____ and _____ 1983. Kin recognition in animals. *Amer. Scient.* 71 : 46-55

Hoogland, J.L. 1982. Prairie dogs avoid extreme inbreeding. *Science* 215 : 1639-41

_____ 1985. Infanticide in prairie dogs : lactating females kill offspring of close kin. *Science* 230 : 1037-40

Hörnfeldt, B. 1978. Synchronous population fluctuations in voles, small game, owls, and tularemia in Northern Sweden. *Oecologia (Berl.)* 32 : 141-52

_____, Löfgren, O. and Carlsson, B.-G. 1986. Cycles in voles and small game in relation to variations in plant production indices in Northern Sweden. *Oecologia*

(Berl.) 68 : 496-502

Horton, T.H. 1984. Growth and maturation in *Microtus montanus*: effects of photoperiods before and after weaning. *Can. J. Zool. 62* : 1741-6

Houston, A.I., McCleery, R.H. and Davies, N.B. 1985. Territory size, prey renewal and feeding rates: interpretation of observations on the pied wagtail (*Motacilla alba*) by simulation. *J. Anim. Ecol. 54* : 227-40

Hrdy, S.B. 1979. Infanticide among animals : a review, classification, and examination of the implications for the reproductive strategies of females. *Ethol. Sociobiol. 1* : 13-40

_____ and Hausfater, G. 1984. Comparative and evolutionary perspectives on infanticide: introduction and overview. In G. Hausfater and S.B. Hrdy (eds.) *Infanticide: Comparative and Evolutionary Perspectives*. Aldine Publishing Company, New York. pp. xiii-xxxv

Huck, U.W. 1984. Infanticide and the evolution of pregnancy block in rodents. In G. Hausfater and S.B. Hrdy (eds.) *Infanticide: Comparative and Evolutionary Perspectives*. Aldine Publishing Company, New York. pp. 349-65

_____ and Banks, E.M. 1979. Behavioral components of individual recognition in the collared lemming (*Dicrostonyx groenlandicus*). *Behav. Ecol. Sociobiol. 6* : 85-90

_____ and _____ 1982a. Differential attraction of females to dominant males: olfactory discrimination and mating preference in the brown lemming (*Lemmus trimucronatus*). *Behav. Ecol. Sociobiol. 11* : 217-22

_____ and _____ 1982b. Male dominance status, female choice and mating success in the brown lemming, *Lemmus trimucronatus*. *Anim. Behav. 30* : 665-75

_____ , Soltis, R.L. and Coopersmith, C.B. 1982. Infanticide in male laboratory mice : effects of social status, prior sexual experience, and basis for discrimination between related and unrelated young. *Anim. Behav. 30* : 1158-65

Hudson, P.J. 1986. The effect of a parasitic nematode on the breeding production of red grouse. *J. Anim. Ecol. 55* : 85-92

Huffaker, C.B. 1958. Experimental studies on predation : dispersion factors and predator-prey oscillations. *Hilgardia 27* : 343-83

Hunt, F. 1982. Regulation of population cycles by genetic feedback. *J. Math. Biol. 13* : 271-82

Hurlbert, S.H. 1984. Pseudoreplication and the design of ecological field experiments. *Ecol. Monogr. 54* : 187-211

Hussell, D.J.T. 1972. Factors affecting clutch size in arctic passerines. *Ecol. Monogr. 42* : 317-64

_____ 1985. Clutch size, daylength, and seasonality of resources: comments on Ashmole's hypothesis. *Auk 102* : 632-4

Hyvärinen, H. and Heikura, K. 1971. Effect of age and seasonal rhythm on the growth patterns of some small mammals in Finland and in Kirkenes, Norway. *J. Zool., Lond. 165* : 545-56

Ims, R.A. 1985. The effect of pregnancy failure on the onset of reproduction in sympatric populations of *Clethrionomys glareolus* and *Clethrionomys rufocanus*. *Ann. Zool. Fennici 22* : 309-12

Innes, D.G.L. 1978. A reexamination of litter size in some North American microtines. *Can. J. Zool. 56* : 1488-96

_____ and Millar, J.S. 1981. Body weight, litter size, and energetics of reproduction in *Clethrionomys gapperi* and *Microtus pennsylvanicus*. *Can. J. Zool. 59* : 785-9

Jamieson, I.G. and Zwickel, F.C. 1983. Dispersal and site fidelity in blue grouse. *Can. J. Zool. 61* : 570-3

Jannett, F.J. 1975. The "hip glands" of *Microtus pennsylvanicus* and *M. longicaudus* (Rodentia: Muridae), voles "without" hip glands. *Syst. Zool. 24* : 171-5

_____ 1978. The density dependent formation of extended maternal families of the montane vole, *Microtus montanus* nanus. *Behav. Ecol. Sociobiol. 3* : 245-63
_____ 1980. Social dynamics of the montane vole, *Microtus montanus*, as a paradigm. *Biologist 62* : 3-19
_____ 1981. Sex ratios in high-density populations of the montane vole, *Microtus montanus*, and the behavior of territorial males. *Behav. Ecol. Sociobiol. 8* : 297-307
_____ 1982. Nesting patterns of adult voles, *Microtus montanus*, in field populations. *J. Mamm. 63* : 495-8
_____ 1984. Reproduction of the montane vole, *Microtus montanus*, in subnivean populations. In J.F. Merritt (ed.) *Winter Ecology of Small Mammals*. Special Publication No. 10, Carnegie Museum of Natural History, Pittsburgh, pp 215-24
Jarman, P.J. 1982. Prospects for interspecific comparison in sociobiology. In Kings' College Sociobiology Group (eds.) *Current Problems in Sociobiology*. Cambridge University Press, Cambridge, pp. 323-42
Jeffreys, A.J., Wilson, V. and Thein, S.L. 1985. Hypervariable 'minisatellite' regions in human DNA. *Nature 314* : 67-73
Jonasson, S., Bryant, J.P., Chapin, F.S. and Andersson, M. 1986. Plant phenols and nutrients in relation to variations in climate and rodent grazing. *Amer. Natur. 128* : 394-408
Jonge, G. de 1983. Aggression and group formation in the voles *Microtus agrestis, M. arvalis* and *Clethrionomys glareolus* in relation to intra- and interspecific competition. *Behaviour 84* : 1-73
Kaczmarski, F. 1966. Bioenergetics of pregnancy and lactation in the bank vole. *Acta theriol. 11* : 409-17
Kaikasulo, A. and Tast, J. 1984. Winter breeding of microtine rodents at Kilpisjärvi, Finnish Lapland. In J.F. Merritt (ed.), *Winter Ecology of Small Mammals*. Special Publication No. 10, Carnegie Museum of Natural History, Pittsburgh, pp. 243-52
Kalela, O. 1957. Regulation of reproduction rate in subarctic populations of the vole, *Clethrionomys glareolus. Ann. Acad. Sci. Fennicae (A IV) 34* : 1-60
_____ 1962. On fluctuations in the numbers of arctic and boreal small rodents as a problem of production biology. *Ann. Acad. Sci. Fennica A IV Biol. 66* : 1-38
_____ , Koponen, T., Lind, E.A., Skáren, U. and Tast, J. 1961. Seasonal change of habitat in the Norwegian lemming, *Lemmus lemmus* (L.). *Ann. Acad. Sci. Fennica A. IV Biol. 55* : 1-75
Kaneko, Y 1978. Seasonal and sexual differences in absolute and relative growth in *Microtus montebelli. Acta Theriol. 23* : 75-98
Karlsson, A.F. 1984. Age-differential homing tendencies in displaced bank voles, *Clethrionomys glareolus. Anim. Behav. 32* : 515-19
Kaufmann, J.H. 1983. On the definitions and functions of dominance and territoriality. *Biol. Rev. 58* : 1-20
Kawata, M. 1985a. Mating system and reproductive success in a spring population of the red-backed vole, *Clethrionomys rufocanus bedfordiae. Oikos 45* : 181-90
_____ 1985b. Sex differences in the spatial distribution of genotypes in the red-backed vole, *Clethrionomys rufocanus bedfordiae. J. Mamm. 66* : 384-7
Keith, L.B. 1963. *Wildlife's Ten-Year Cycle*. University of Wisconsin Press, Madison
_____ 1974. Some features of population dynamics in mammals. *Proc. Int. Congr. Game Biol. 11* : 17-58
_____ , Cary, J.R., Rongstad, O.J. and Brittingham, M.C. 1985. Demography and ecology of a declining snowshoe hare population. *Wildl. Monogr. 90* : 1-43
_____ and Windberg, L. 1978. A demographic analysis of the snowshoe hare cycle. *Wildlife Monographs 58* : 1-70

Keith, T.P. and Tamarin, R.H. 1981. Genetic and demographic differences between dispersers and residents in cycling and noncycling vole populations. *J. Mamm.* 62 : 713-25

Keller, B.L. 1985. Reproductive patterns. In R.H. Tamarin (ed.), *Biology of New World Microtus*. Special Publication No. 8 , American Society of Mammalogists, pp. 725-78

_____ and Krebs, C.J. 1970. Microtus population biology III. Reproductive changes in fluctuating populations of *M. ochrogaster* and *M. pennsylvanicus* in southern Indiana, 1965-7. *Ecol. Monogr. 40* : 263-94

Kenagy, G.J. and Hoyt, D.F. 1980. Reingestion of feces in rodents and its daily rhythmicity. *Oecologia (Berl.) 44* : 403-9

Kenney, A.M. and Dewsbury, D.A. 1977. Effect of limited mating on the corpora lutea in montane voles, *Microtus montanus*. *J. Reprod. Fertil. 49* : 363-4

_____ , Evans, R.L. and Dewsbury, D.A. 1977. Postimplantation pregnancy disruption in *Microtus ochrogaster, M. pennsylvanicus* and *Peromyscus maniculatus*. *J. Reprod. Fertil 49* : 365-7

_____ , Hartung, T.G. and Dewsbury, D.A. 1979. Copulatory behavior and the initiation of pregnancy in California voles (*Microtus californicus*). *Brain Behav. Evol. 16* : 176-91

_____ , Lanier, D.L. and Dewsbury, D.A. 1977. Effects of vaginal-cervical stimulation in seven species of muroid rodents. *J. Reprod. Fertil. 49* : 305-9

Keppie, D.M. 1979. Dispersal, overwinter mortality, and recruitment of spruce grouse. *J. Wildl. Manage. 43* : 717-27

_____ 1980. Similarities of dispersal among sibling male spruce grouse. *Can. J. Zool. 58* : 2102-4

Kessler, L.G. and Avise, J.C. 1985. Microgeographic lineage analysis by mitochondrial genotype: variation in the cotton rat (*Sigmodon hispidus*). *Evolution 39* : 831-37

Keverne, E.B. 1983. Pheromonal influences on the endocrine regulation of reproduction. *Trends Neurosc. 8* : 381-4

Kimura, M. 1983. *The Neutral Theory of Molecular Evolution.* Cambridge University Press, Cambridge.

Kleiman, D.G. 1977. Monogamy in mammals. *Q. Rev. Biol. 52* : 39-69

_____ 1981. Correlations among life history characteristics of mammalian species exhiniting two extreme forms of monogamy. In R.D. Alexander and D.W. Tinkle (eds.) *Natural Selection and Social Behavior: Recent Research and New Theory.* Chiron Press, New York, pp. 332-44

Klevezal, G.A., Pucek, M. and Malafeeva, E.P. 1984. Body and skeleton growth in laboratory field voles of different seasonal generations. *Acta Theriol. 29* : 3-16

Klomp, H. 1970. The determination of clutch-size in birds: a review. *Ardea 58* : 1-124

Koenig, W.D. 1984. Geographic variation in clutch size in the northern flicker (*Colaptes auratus*): support for Ashmole's hypothesis. *Auk 101* : 698-706

_____ , Mumme, R.L. and Pitelka, F.A. 1984. The breeding system of the acorn woodpecker in central coastal California. *Z. Tierpsychol. 65* : 289-308

Kohn, P.H. and Tamarin, R.H. 1978. Selection at electrophoretic loci for reproductive parameters in island and mainland voles. *Evolution 32* : 15-28

Korn, H. 1986. Changes in home range size during growth and maturation of the wood mouse (*Apodemus sylvaticus*) and the bank vole (*Clethrionomys glareolus*). *Oecologia (Berl.) 68* : 623-8

Korpimäki, E. 1985. Rapid tracking of microtine populations by their avian predators: possible evidence for stabilizing predation. *Oikos 45* : 281-4

_____ 1986. Predation causing synchronous decline phases in microtine and shrew populations in western Finland. *Oikos 46* : 124-7

Koshkina, T.V. and Korotkov, Y.S. 1975. Regulative adaptations in populations of the red vole (*Clethrionomys rutilus*) under optimum conditions of its range [in Russian; translation by W.A. Fuller]. *Fauna Ecol. Rodents 12* : 5-61

Kot, M. and Schaffer, W.M. 1984. The effects of seasonality on discrete models of population growth. *Theor. Popul. Biol. 26* : 340-60

Krebs, C.J. 1964a. The lemming cycle at Baker Lake, Northwest Territories, during 1959-1967. *Tech. Paper Arctic Inst. N. Amer. 15* : 1-104

_____ 1964b. Cyclic variation in skull-body regressions of lemmings. *Can. J. Zool. 42* : 631-43

_____ 1966. Demographic changes in fluctuating populations of Microtus californicus. *Ecol. Monogr. 36* : 239-73

_____ 1970. *Microtus* population biology: behavioral changes associated with the population cycle in *M. ochrogaster* and *M. pennsylvanicus*. *Ecology 51* : 34-52

_____ 1978. A review of the Chitty hypothesis of population regulation. *Can. J. Zool. 56* : 2463-80

_____ 1979. Dispersal, spacing behaviour, and genetics in relation to population fluctuations in the vole *Microtus townsendii*. *Fortschr. Zool. 25* : 61-77

_____ 1985a. *Ecology : the Experimental Analysis of Distribution and Abundance*. 3rd edn. Harper and Row, New York.

_____ 1985b. Do changes in spacing behaviour drive population cycles in small mammals? In R.M. Sibly and R.H. Smith (eds.) *Behavioural Ecology: Ecological Consequences of Adaptive Behaviour*. Blackwell Scientific Publications, Oxford, pp. 295-312

_____ 1987. The role of dispersal in cyclic rodent populations. In N.C. Stenseth and W.Z. Lidicker (eds.). *Animal Dispersal : Small Mammals as a Model*. Chapman and Hall, London, pp. 000-000

_____ and Boonstra, R. 1978. Demography of the spring decline in populations of the vole, *Microtus townsendii*. *J. Anim. Ecol. 47* : 1007-15

_____ and DeLong, K.T. 1965. A *Microtus* population with supplemental food. *J. Mamm. 46* : 566-73

_____, Halpin, Z.T. and Smith, J.N.M. 1977. Aggression, testosterone and the spring decline in populations of the vole, *Microtus townsendii*. *Can. J. Zool. 55* : 430-7

_____, Keller, B.L. and Tamarin, R.H. 1969. *Microtus* population biology: demographic changes in fluctuating populations of *Microtus ochrogaster* and *M. pennsylvanicus* in southern Indiana. *Ecology 50* : 587-607

_____ and Myers, J.H. 1974. Population cycles in small mammals. *Adv. Ecol. Res. 8* : 267-399

_____, Wingate, I., LeDuc, J., Redfield, J.A., Taitt, M. and Hilborn, R. 1976. *Microtus* population biology: dispersal in fluctuating populations of *M. townsendii*. *Can. J. Zool. 54* : 79-95

Krebs, J.R. 1971. Territory and breeding density in the great tit, Parus major. *Ecology 52* : 2-22

_____ 1982. Territorial defence in the Great Tit (*Parus major*): Do residents always win? *Behav. Ecol. Sociobiol. 11* : 185-94

_____ 1985. Sociobiology ten years on. *New Scientist 1476* : 40-3

_____ and Dawkins, R. 1984. Animal signals: mind-reading and manipulation. In J.R. Krebs and N.B. Davies (eds.) *Behavioural Ecology : An Evolutionary Approach*. 2nd ed. Blackwell Scientific Publications, Oxford, pp. 380-402

Krohne, D.T. 1980. Intraspecific litter size variation in *Microtus californicus*. II. Variation between populations. *Evolution 34* : 1174-82

_____ 1981. Intraspecific litter size variation in *Microtus californicus*: variation within populations. *J. Mamm. 62* : 29-40

_____ and Baccus, R. 1985. Genetic and ecological structure of a population of *Peromyscus leucopus*. *J. Mamm.* 66 : 529-37

_____, Dubbs, B.A. and Baccus, R. 1984. Analysis of dispersal in an unmanipulated population of *Peromyscus leucopus*. *Amer. Midl. Natur.* 112 : 146-56

_____ and Miner, M.S. 1985. Removal trapping studies of dispersal in *Peromyscus leucopus*. *Can. J. Zool.* 63 : 71-5

Kruczek, M. 1986. Seasonal effects on sexual maturation of male bank voles (*Clethrionomys glareolus*). *J. Reprod. Fertil.* 76 : 83-9

Kuno, E. 1981. Dispersal and the persistence of populations in unstable habitats : a theoretical note. *Oecologia (Berl.)* 49 : 123-6

Labov, J. 1980. Factors influencing infanticidal behavior in wild house mice (*Mus musculus*). *Behav. Ecol. Sociobiol.* 6 : 297-303

_____ 1981a. Pregnancy blocking in rodents: adaptive advantages for females. *Amer. Natur.* 118 : 361-71

_____ 1981b. Male social status, physiology, and ability to block pregnancies in female house mice (*Mus musculus*). *Behav. Ecol. Sociobiol.* 8 : 287-91

_____, Huck, U.W., Elwood, R.E. and Brooks, R.J. 1985. Current problems in the study of infanticidal behavior of rodents. *Q. Rev. Biol.* 60 : 1-20

_____, _____, Vaswani, P. and Lisk, R.D. 1986. Sex ratio manipulation and decreased growth of male offspring of undernourished golden hamsters (*Mesocricetus auratus*). *Behav. Ecol. Sociobiol.* 18 : 241-8

Lack, D. 1947. The significance of clutch size. *Ibis 89* : 302-52

_____ 1954. *The Natural Regulation of Animal Numbers*. Clarendon Press, Oxford

_____ 1968. *Ecological Adaptations for Breeding in Birds*. Methuen, London.

Lacy, R.C. 1978. Dynamics of t-alleles in *Mus musculus* populations: review and speculation. *Biologist* 60 : 41-67

Laine, K. and Hentonnen, H. 1983. The role of plant production in microtine cycles in northern Fennoscandia. *Oikos* 40 : 407-18

Lance, A.N. 1983. Selection of feeding sites by hen red grouse *Lagopus lagopus scoticus* during breeding. *Ornis Scandinavica 14* : 78-80

Lande, R. 1981. Models of speciation by sexual selection on polygenic traits. *Proc. Natl. Acad. Sci. USA* 78 : 3721-5

_____ 1982. A quantitative genetic theory of life history evolution. *Ecology 63* : 607-15

Larsson, T.-B. 1977. Small rodent abundance in relation to reforestation measures and natural habitat variables in northern Sweden. *EPPO Bull.* 7 : 397-409

Lawton, A.D. and Whitsett, J.M. 1979. Inhibition of sexual maturation by a urinary pheromone in male prairie deermice. *Horm. Behav. 13* : 128-35

Lecyk, M. 1967. Influence of crowded population stimuli on the development of the reproductive organs in the common vole. *Acta theriol. 12* : 177-9

Lee, A.K., Bradley, A.J. and Braithwaite, R.W. 1977. Corticosteroid levels and male mortality in *Antechinus stuartii*. In B. Stonehouse and D. Gilmore (eds.) *The Biology of Marsupials*. Macmillan, London, pp. 209-20

_____ and Cockburn, A. 1985a. *Evolutionary Ecology of Marsupials*. Cambridge University Press, Cambridge

_____ and _____ 1985b. Spring declines in small mammal populations. Acta Zool. Fennica 173 : 75-6

_____ and McDonald, I.R. 1985. Stress and population regulation in small mammals. *Oxford Rev. Reprod. Biol.* 7 : 261-304

Lefkovitch, L.P. and Fahrig, L. 1985. Spatial characteristics of habitat patches and population survival. *Ecological Modelling 30* : 297-308

Lendrem, D.W. 1985. Kinship accelerates puberty acceleration in mice (*Mus musculus*).

Behav. Ecol. Sociobiol. 17 : 397-9

Lenington, S. 1983. Social preferences for partners carrying 'good genes' in wild mice. *Anim. Behav. 31* : 325-33

_____ and Egid, K. 1985. Female discrimination of male odors correlated with male genotype at the T locus in *Mus musculus* : a response to T-locus or H-2- locus variability? *Behav. Genet. 15* : 53-67

Lepri, J.J. and Noden, P.F. 1984. Reproductive function is independent of photoperiod in adult male *Microtus pinetorum*. *J. Mamm. 65* : 696-7

Leslie, P.H., Tener, T.S., Vizoso, M. and Chitty, H. 1955. The longevity and fertility of the Orkney vole, *Microtus orcadensis*, as observed in the laboratory. *Proc. Zool. Soc. Lond. 125* : 115-25

Levin, S.A. 1976. Population dynamics in heterogeneous environments. *Ann. Rev. Ecol. Syst. 7* : 287-310

_____ , Cohen, D. and Hastings, A. 1984. Dispersal strategies in patchy environments. *Theor. Popul. Biol. 26* : 165-91

Levins, R. 1968. *Evolution in Changing Environments.* Princeton University Press, Princeton

Lewin, R. 1984. Practice catches theory in kin recognition. *Science 223* : 1049-51

Lewis, R.A. 1984. Density, movements, and breeding success of female blue grouse in an area of reduced male density. *Can. J. Zool. 62* : 1556-60

_____ and Zwickel, F.C. 1980. Removal and persistence of male blue grouse on persistent and transient territorial sites. *Can. J. Zool. 58* : 1417-23

Lewontin, R.C. 1974. *The Genetic Basis of Evolutionary Change.* Columbia University Press, Columbia.

_____ 1985. Population genetics. In P.J. Greenwood, P.H. Harvey and M. Slatkin (eds.) *Evolution : Essays in Honour of John Maynard Smith.* Cambridge University Press, Cambridge, pp. 3-18

_____ and Krakauer, J. 1973. Distribution of gene frequency as a test of the theory of the selective neutrality of polymorphisms. *Genetics 74* : 175-95

Liapunova, E.A., Vorontsov, N.N. and Zakaran, G.G. 1975. Zygotic mortality in *Ellobius lutescens* (Rodentia, Microtinae). *Experientia 31 :* 417-8

Liberg, O. and Von Schantz, T. 1985. Sex-biased philopatry and dispersal in birds and mammals: the Oedipus hypothesis. *Amer. Natur. 126* : 129-35

Lidicker, W.Z. 1973. Regulation of numbers in an island population of the California vole, a problem in community dynamics. *Ecol. Monogr. 43* : 271-302

_____ 1975. The role of dispersal in the demography of small mammals. In K. Petrusewicz, F.B. Golley and L. Ryszkowski (eds.) *Small Mammals: Productivity and Dynamics of Populations.* Cambridge University Press, Cambridge, pp. 103-28

_____ 1976. Experimental manipulation of the timing of reproduction in the California vole. *Res. Popul. Ecol. 18* : 14-27

_____ 1978. Regulation of numbers in small mammal populations - historical reflections and a synthesis. In D.P Snyder (ed.) *Populations of Small Mammals under Natural Conditions.* Pymatuning Laboratory of Ecology, Pittsburgh, pp. 122-41

_____ 1979. Analysis of two freely-growing enclosed populations of the California vole. *J. Mamm. 60* : 447-66

_____ 1980. The social biology of the California vole. *The Biologist 62* : 46-55

_____ 1985a. Population structuring as a factor in understanding microtine cycles. *Acta Zool. Fennica 173* : 23-7

_____ 1985b. Dispersal. In R.H. Tamarin (ed.), *Biology of New World* Microtus. Special Publication No. 8 , American Society of Mammalogists, pp. 420-54

Lima, S. 1984. Territoriality in variable environments : a simple model. *Amer. Natur. 124* : 641-55

Lindroth, R.L. and Batzli, G.O. 1984. Food habits of the meadow vole (*Microtus pennsylvanicus*) in bluegrass and prairie habitats. *J. Mamm.* 65 : 139-43
_____ and _____ 1986. Inducible plant chemical defences: a cause of vole population cycles? *J. Anim. Ecol.* 55 : 431-49
Linzey, A.V. 1984. Patterns of coexistence in *Synaptomys cooperi* and *Microtus pennsylvanicus*. *Ecology* 65 : 382-93
Livdahl, T.P. 1979. Environmental uncertainty and selection for life-cycle delays in opportunistic species. *Amer. Natur.* 113 : 835-42
Lizzaralde, M.S., Bianchi, N.O. and Merani, M.S. 1982. Cytogenetics of South American akodont rodents (Cricetidae). VII. Origin of sex ratio polymorphism in *Akodon azarae*. *Cytologia* 47 : 183-93
Lobato, L., Cantos, G., Araujo, B., Bianchi, N.O. and Merani, M.S. 1982. Cytogenetics of South American akodont rodents (Cricetidae). X. *Akodon mollis*: a species with XY females and B chromosomes. *Genetica* 57 : 199-205
Lochmiller, R.L., Whelan, J.B. and Kirkpatrick, R.L. 1982. Energetic cost of lactation in *Microtus pinetorum*. *J. Mamm.* 63 : 475-81
_____, _____ and _____ 1983. Seasonal energy requirements of adult pine voles, *Microtus pinetorum*. *J. Mamm.* 64 : 345-50
Lombardo, D.L. and Terman, C.R. 1980. The influence of the social environment on sexual maturation of female deermice (*Peromyscus maniculatus bairdii*). *Res. Popul. Ecol.* 27 : 93-100
Lomnicki, A. 1978. Individual differences between animals and natural regulation of their numbers. *J. Anim. Ecol.* 47 : 461-75
_____ 1980. Regulation of population density due to individual differences and patchy environment. *Oikos* 35 : 185-93
_____ 1982. Individual heterogeneity and population regulation. In Kings' College Sociobiology Group (eds.) *Current Problems in Sociobiology*. Cambridge University Press, Cambridge, pp. 153-67
_____ and Ombach, J. 1984. Resource partitioning within a single species population and population stability : a theoretical model. *Theor. Popul. Biol.* 24 : 21-8
Lomolino, M.V. 1984. Immigrant selection, predation, and the distributions of *Microtus pennsylvanicus* and *Blarina brevicauda* on islands. *Amer. Natur.* 123 : 468-83
Lord, R.D. 1960. Litter size and latitude in North American mammals. *Amer. Midl. Natur.* 64 : 488-99
Ludwig, D. 1984. *Microtus richardsoni* microhabitat and life history. In J.F. Merritt (ed.) *Winter Ecology of Small Mammals*. Special Publication No. 10, Carnegie Museum of Natural History, Pittsburgh, pp. 319-31
Lynch, M. 1984. The limits to life history evolution in *Daphnia*. *Evolution* 38 : 465-82
MacArthur, R.H. and Wilson, E.O. 1967. *The Theory of Island Biogeography*. Princeton University Press, Princeton
McCleery, R.H. and Perrins, C.M. 1985. Territory size, reproductive success and poulation dynamics in the great tit, *Parus major*. In R.M. Sibly and R.H. Smith (eds.) *Behavioural Ecology: Ecological Consequences of Adaptive Behaviour*. Blackwell Scientific Publications, Oxford, pp. 353-73
McClure, P.A. 1981. Sex-biased litter reduction in food-restricted wood rats (*Neotoma floridana*). *Science* 211 : 1058-60
McDonald, I.R., Lee, A.K., Than, K.A. and Martin, R.W. 1986. Failure of glucocorticoid feedback in males of a population of small marsupials (*Antechinus swainsonii*) during the period of mating. *J. Endocrin.* 108 : 63-8
_____ and Taitt, M.J. 1982. Steroid hormones in the blood plasma of Townsend's vole (*Microtus townsendii*). *Can. J. Zool.* 60 : 2264-9

McGovern, M. and Tracy, C.R. 1981. Phenotypic variation in electromorphs previously considered to be genetic markers in *Microtus ochrogaster*. *Oecologia (Berl.) 51* : 276-80

_____ and _____ 1985. Physiological plasticity in electromorphs of blood proteins in free-ranging *Microtus ochrogaster*. *Ecology 66* : 396-403

McGregor, P.K. and Krebs, J.R. 1982. Mating and song types in the great tit. *Nature 297* : 60-61

McGuire, B. and Novak, M. 1984. A comparison of maternal behaviour in the meadow vole (*Microtus pennsylvanicus*), prairie vole (*M. ochrogaster*) and pine vole (*M. pinetorum*). *Anim. Behav. 32* : 1132-41

McGuire, M.R. and Getz, L.L. 1981. Incest taboo between sibling *Microtus ochrogaster*. *J. Mamm. 62* : 213-15

McIntosh, R.P. 1980. The background and some current problems of theoretical ecology. *Synthese 43* : 195-255

_____ 1985. *The Background of Ecology: Concept and Theory*. Cambridge University Press, Cambridge

McNab, B.K. 1980. Food habits, energetics, and the population biology of mammals. *Amer. Natur. 116* : 106-24

McNaughton, S.J. 1979. Grazing as an optimization process : grass-ungulate relationships in the Serengeti. *Amer. Natur. 113* : 691-705

McShea, W.J. and Madison, D.M. 1984. Communal nesting between reproductively active females in a spring population of *Microtus pennsylvanicus*. *Can. J. Zool. 62* : 344-6

_____ and _____ 1986. Sex ratio shifts within litters of meadow voles (*Microtus pennsylvanicus*). *Behav. Ecol. Sociobiol. 18* : 431-6

Mace, G.M. and Harvey, P.H. 1983. Energetic constraints on home-range size. *Amer. Natur. 121* : 120-32

Mackin-Rogalska, R. 1979. Elements of the spatial organisation of a common vole population. *Acta theriol. 24* : 171-200

Madison, D.M. 1978. Movement indicators of reproductive events among female meadow voles as revealed by radiotelemetry. *J. Mamm. 59* : 835-43

_____ 1980a. Space use and social structure in meadow voles, *Microtus pennsylvanicus*. *Behav. Ecol. Sociobiol. 7* : 65-71

_____ 1980b. An integrated view of the social biology of *Microtus pennsylvanicus*. *The Biologist 62* : 20-33

_____ 1981. Time patterning of nest visitation by lactating meadow voles. *J. Mamm. 62* : 389-90

_____ 1984. Group nesting and its ecological and evolutionary significance in overwintering microtine rodents. In J.F. Merritt (ed.) *Winter Ecology of Small Mammals*. Special Publication No. 10, Carnegie Museum of Natural History, Pittsburgh, pp. 267-74

_____ 1985. Activity rhythms and spacing. In R.H. Tamarin (ed.) *Biology of New World* Microtus. Special Publication No. 8, American Society of Mammalogists, pp. 373-419

_____, FitzGerald, R.W. and McShea, W.J. 1984. Dynamics of social nesting in overwintering meadow voles (*Microtus pennsylvanicus*): possible consequences for population cycling. *Behav. Ecol. Sociobiol. 15* : 9-17

Mallory, F.F. and Brooks, R.J. 1978. Infanticide and other reproductive strategies in the collared lemming, *Dicrostonyx groenlandicus*. *Nature 273* : 144-6

_____ and _____ 1980. Infanticide and pregnancy failure: reproductive strategies in the female collared lemming (*Dicrostonyx groenlandicus*). *Biol. Reprod. 22* : 192-6

_____ , _____ and Elliott, J.R. 1986. Variations of skull-body regressions of the

lemming (*Dicrostonyx groenlandicus*) under laboratory and field conditions. *Zool. J. Linn. Soc.* 87 : 125-38
—— and Clulow, F.V. 1977. Evidence of pregnancy failure in the wild meadow vole, *Microtus pennsylvanicus. Can. J. Zool.* 55 : 1-17
——, Elliott, J.R. and Brooks, R.J. 1981. Changes in body size in fluctuating populations of the collared lemming: age and photoperiod influences. *Can. J. Zool.* 59 : 174-82
Malzahn, E. 1981. Trace elements and their significance in the postnatal development of seasonal generations of the bank vole. *Acta theriol.* 26 : 231-56
—— 1983. Postnatal changes in trace elements and in oxidation-reduction activity in laboratory bank voles. *Acta theriol.* 28 : 33-54
Manning, J.T. 1985. Choosy females and correlates of male age. *J. theor. Biol.* 116 : 349-54
Martin. K. 1984. Reproductive defence priorities of male willow ptarmigan (*Lagopus lagopus*): enhancing mate survival or extending paternity options? *Behav. Ecol. Sociobiol.* 16 : 57-63
Martinet, L. and Spitz, F. 1971. Variation saisonnieres de la croissance et de la mortalité du campagnol des champs, *Microtus arvalis*. Rôle du photopériodisme et de la végétation sur ces variations. *Mammalia* 35 : 38-84
Maruniak, J.A., Coquelin, A. and Bronson, F.H. 1978. The release of LH in male mice in response to female urinary odors: characteristics of the response in young males. *Biol. Reprod.* 18 : 251-5
Maser, C. 1974. The sage vole, *Lagurus curtatus* (Cope, 1868), in the Crooked River National Grassland, Jefferson County, Oregon. A contribution to its life history and ecology. *Säuget. Mitteil.* 22 : 193-222
Massey, A. and Vandenbergh, J.G. 1980. Puberty delay by a urinary cue from female house mice in feral populations. *Science* 209 : 821-2
—— and —— 1981. Puberty acceleration by a urinary cue from male mice in feral populations. *Biol. Reprod.* 24 : 523-7
Mattingly, D.K. and McClure, P.A. 1982. Energetics of reproduction in large-littered cotton rats, *Sigmodon hispidus. Ecol. Monogr.* 63 : 183-95
—— and —— 1985. Energy allocation during lactation in cotton rats (*Sigmodon hispidus*) on a restricted diet. *Ecology* 66 : 928-37
May, R.M. 1974. Biological populations with non-overlapping generations: stable points, stable cycles and chaos. *Science* 186 : 645-7
—— 1976. Simple mathematical models with very complicated dynamics. *Nature* 261 : 459-67
—— 1981a. Models for single populations. In R.M. May (ed.) *Theoretical Ecology: Principles and Applications.* 2nd ed. Blackwell Scientific Publications, Oxford, pp. 5-29
—— 1981b. Models for two interacting populations. In R.M. May (ed.) *Theoretical Ecology: Principles and Applications.* 2nd ed. Blackwell Scientific Publications, Oxford, pp. 78-104
—— 1986. The cautionary tale of the black-footed ferret. *Nature* 320 : 13-14
—— and Anderson, R.M. 1978. Regulation and stability of host-parasite population interactions. I. Destabilizing processes. *J. Anim. Ecol.* 47 : 249-67
—— and Oster, G.F. 1976. Bifurcations and dynamic complexity in simple ecological models. *Amer. Natur.* 110 : 573-99
Maynard Smith, J. 1964. Group selection and kin selection. *Nature* 201 : 1145-7
—— 1974. *Models in Ecology.* Cambridge University Press, Cambridge
—— 1980. A new theory of sexual investment. *Behav. Ecol. Sociobiol.* 7 : 247-51
—— 1982. *Evolution and the Theory of Games.* Cambridge University Press,

Cambridge
_____ 1983. The population as a unit of selection. In B. Shorrocks (ed.) *Evolutionary Ecology*. Blackwell Scientific Publications, Oxford, pp. 195-202
_____ 1985a. Sexual selection, handicaps and true fitness. *J. theor. Biol. 115* : 1-8
_____ 1985b. Appendix. In R.M. Sibly and R.H. Smith (ed.) *Behavioural Ecology: Ecological Consequences of Adaptive Behaviour*. Blackwell Scientific Publications, Oxford, pp. 72-3
_____ and Price, G.R. 1973. The logic of animal conflict. *Nature 246* : 15-18
_____ and Stenseth, N.C. 1978. On the evolutionary stability of the female-biased sex ratio in the wood lemming (*Myopus schisticolor*): the effect of inbreeding. *Heredity 41* : 205-14
Mazurkiewicz, M. 1971. Shape, size and distribution of home ranges of *Clethrionomys glareolus* (Schreber, 1780). *Acta theriol. 16* : 23-60
_____ 1981. Spatial organization of a bank vole population in years of small or large numbers. *Acta theriol. 26* : 31-45
_____ and Rajska, E. 1975. Dispersion of young bank voles from their place of birth. *Acta theriol 16* : 23-60
Merritt, J.F. (ed.)1984. *Winter Ecology of Small Mammals*. Carnegie Museum of Natural History. Special Publication No.10, Pittsburgh
Mertz, D.B. 1971. Life history phenomena in increasing and decreasing populations. In G.P. Patil, E.G. Pielou and W.E. Waters (eds.) *Symposium on Statistical Ecology, Vol. 2*. Penn State University Press, University Park, Pennsylvania, pp. 351-400
_____ , Craig, D.M., Wade, M.J. and Boyer, J.F. 1984. Cohort selection. *Evolution 38* : 560-70
Migula, P. 1969. Bioenergetics of pregnancy and lactation in the European common vole. *Acta theriol. 14* : 167-79
Mihok, S. 1979. Behavioral strcuture and demography of subarctic *Clethrionomy gapperi* and *Peromyscus maniculatus*. *Can. J. Zool. 57* : 1520-35
_____ 1981. Chitty's hypothesis and behaviour in subarctic red-backed voles *Clethrionomys gapperi*. *Oikos 36* : 281-95
_____ and Ewing, D. 1983. Reliability of transferrin and leucine aminopeptidase phenotyping in wild meadow voles (*Microtus pennsylvanicus*). *Biochemical Genetics 21* : 969-83
_____ , Fuller, W.A., Canham, R.P. and McPhee, E.C. 1983. Genetic changes at the transferrin locus in the red-backed vole (*Clethrionomys gapperi*). *Evolution 37* : 332-40
Millar, J.S. 1982. Life cycle characteristics of northern *Peromyscus maniculatus borealis*. *Can. J. Zool. 60* : 510-5
_____ and Innes, D.G.L. 1983. Demographic and seasonal characteristics of montane deer mice. *Can. J. Zool. 61* : 574-85
_____ , Wille, F.B. and Iverson, S.L. 1979. Breeding by *Peromyscus* in seasonal environments. *Can. J. Zool. 57* : 719-27
Miller, G.R., Watson, A. and Jenkins, D. 1970. Responses of red grouse to experimental improvement of their food. In A. Watson (ed.) *Animal Populations in Relation to their Food Resources*. Blackwell Scientific Publications, Oxford, pp. 323-34
Milligan, S.R. 1975. The copulatory pattern of *Microtus agrestis*. *J. Mamm. 56* : 220-4
_____ 1979. The copulatory pattern of the bank vole (*Clethrionomys glareolus*) and speculation on the role of penile spines. *J. Zool. Lond. 188* : 279-283
_____ , Charlton, H.M. and Versi, E. 1979. Evidence for a coitally controlled 'mnemonic' involved in luteal function in the vole (*Microtus agrestis*). *J. Reprod. Fertil. 57* : 227-33

Millington, S.J. and Price, T.D. 1985. Song inheritance and mating patterns in Darwin's finches. *Auk 102* : 342-6

Moore, J. and Ali, R. 1984. Are dispersal and inbreeding avoidance related? *Anim. Behav. 32* : 94-112

Morris, D.W. 1984. Rodent population cycles: life history adjustments to age-specific dispersal strategies and intrinsic time lags. *Oecologia (Berl.) 64* : 8-13

———— 1985. Natural selection for reproductive optima. *Oikos 45* : 290-2

———— 1986. Proximate and ultimate controls on life-history variation: the evolution of litter size in white-footed mice (*Peromyscus leucopus*). *Evolution 40* : 169-81

Morrison, P.R., Dieterich, R. and Preston, D. 1977. Body growth in sixteen rodent species and subspecies maintained in laboratory colonies. *Physiol. Zool. 50* : 294-310

Moss, R. 1972. Food selection by red grouse (*Lagopus lagopus scoticus* (Lath.)) in relation to chemical composition. *J. Anim. Ecol. 41* : 411-8

———— 1983. Gut size, body weight, and digestion of winter foods by grouse and ptarmigan. *Condor 85* : 185-93

———— and Hewson, R. 1985. Effects on heather of heavy grazing by mountain hares. *Holarctic Ecology 8* : 280-94

————, Kolb, H.H., Marquiss, M., Watson, A., Tréca, B., Watt, D. and Glennie, W. 1979. Aggressiveness and dominance in captive cock red grouse. *Aggressive Behaviour 5* : 89-94

———— and Miller, G.R. 1976. Production, dieback and grazing of heather (*Calluna vulgaris*) in relation to numbers of red grouse (*Lagopus l. scoticus*) and mountain hares (*Lepus timidus*) in north-east Scotland. *J. appl. Ecol. 13* : 369-77

————, ———— and Allen, S.E. 1972. Selection of heather by captive red grouse in relation to the age of the plant. *J. appl. Ecol. 9* : 771-81

————, Rothery, P. and Trenholm, I.B. 1985. The inheritance of social dominance rank in red grouse (*Lagopus lagopus scoticus*). *Aggressive Behaviour 11* : 257-9

———— and Watson, A. 1980. Inherent changes in the aggressive behaviour of a fluctuating red grouse *Lagopus lagopus scoticus* population. *Ardea 68* : 113-9

———— and ———— 1982. Heritability of egg size, hatch weight, body weight, and viability in red grouse (*Lagopus lagopus scoticus*). *Auk 99* : 683-6

———— and ———— 1984. Maternal nutrition, egg quality and breeding success of Scottish ptarmigan *Lagopus mutus*. *Ibis 126* : 212-20

———— and ———— 1985. Adaptive value of spacing behaviour in population cycles of red grouse and other animals. In R.M. Sibly and R.H. Smith (eds.) *Behavioural Ecology: Ecological Consequences of Adaptive Behaviour.* Blackwell Scientific Publications, Oxford, pp. 275-94

————, ———— and Rothery, P. 1984. Inherent changes in body size, viability and behaviour of a fluctuating red grouse (*Lagopus lagopus scoticus*) population. *J. Anim. Ecol. 53* : 171-89

————, ————, ———— and Glennie, W.W. 1982. Inheritance of dominance and aggressiveness in captive red grouse *Lagopus lagopus scoticus*. *Aggressive Behaviour 8* : 1-18

Motro, U. 1982a. Optimal rates of dispersal. I. Haploid populations. *Theor. Popul. Biol. 21* : 394-411

———— 1982b. Optimal rates of dispersal. II. Diploid populations. *Theor. Popul. Biol. 21* : 412-29

———— 1983. Optimal rates of dispersal. III. Parent-offspring conflict. *Theor. Popul. Biol. 23* : 159-68

———— and Thomson, G. 1982. On heterozygosity and the effective size of populations subject to size changes. *Evolution 36* : 1059-66

Mueller, L.D., Barr, L.G. and Ayala, F.J. 1985. Natural selection vs random drift: evidence from temporal variation in allele frequencies in nature. *Genetics 111* : 517-57

Mullen, D.A. and Pitelka, F.A. 1972. Efficiency of winter scavengers in the Arctic. *Arctic 25* : 225-31

Mullican, T.R. and Keller, B.L. 1986. Ecology of the sagebrush vole (*Lemniscus curtatus*) in southeastern Idaho. *Can. J. Zool. 64* : 1218-23

Munn, C.A. 1986. Birds that 'cry wolf'. *Nature 319* : 143-5

Murphy, D.D., Launer, A.E. and Ehrlich, P.R. 1983. The role of adult feeding in egg production and population dynamics of the Checkerspot butterfly, *Euphydras editha*. *Oecologia (Berl.) 56* : 257-63

Murphy, G.I. 1968. Pattern of life history and the environment. *Amer. Natur. 102* : 390-404

Murray, B.G. 1984. A demographic theory on the evolution of mating systems as exemplified by birds. *Evol. Biol. 18* : 71-140

Myers, J.H. and Krebs, C.J. 1971a. Sex ratios in open and enclosed vole populations: demographic implications. *Amer. Natur. 105* : 325-344

_____ and _____ 1971b. Genetic, behavioural, and reproductive attributes of dispersing field voles *Microtus pennsylvanicus* and *Microtus ochrogaster*. *Ecol. Monogr. 41* : 53-78

Myers, P. and Master, L.L. 1983. Reproduction by *Peromyscus maniculatus*: size and compromise. *J. Mamm. 64* : 1-8

_____ , _____ and Garrett, R.A. 1985. Ambient temperature and rainfall: an effect on sex ratio and litter size in deer mice. *J. Mamm. 66* : 289-98

Myllymäki, A. 1977a. Demographic mechanisms in the fluctuating populations of the field vole *Microtus agrestis*. *Oikos 29* : 468-93

_____ 1977b. Intraspecific competition and home range dynamics in the field vole *Microtus agrestis*. *Oikos 29* : 553-69

Nabaglo, L. 1981. Demographic processes in a confined population of the common vole. *Acta theriol. 26* : 163-83

Nakatsura, K. and Kramer, D.L. 1982. Is sperm cheap? Limited male fertility and female choice in the Lemon Tetra (Pisces, Characidae). *Science 216* : 753-5

Naumov, N.P. 1954. *The Ecology of Mammals* [in Russian]. Sov Nauka, Moscow

_____ 1964. Spatial features and mechanisms of the dynamics of numbers in terrestrial vertebrates. In *Contemporary Problems in the Study of the Dynamics of Numbers in Animal Populations* [in Russian]. Izd Akad Nauk SSR, Moscow

Negus, N.C. and Berger, P.J. 1977. Experimental triggering of reproduction in a natural population of *Microtus montanus*. *Science 196* : 1230-1

_____ and Pinter, A.J. 1965. Litter sizes of *Microtus montanus* in the laboratory. *J. Mamm. 46* : 434-7

Nelson, R.J., Dark, J. and Zucker, I. 1983. Influence of photoperiod, nutrition and water availability on reproduction of male California voles (*Microtus californicus*). *J. Reprod. Fertil. 69* : 473-7

Newsome, A.E. 1969a. A population study of house mice temporarily inhabiting a South Australian wheatfield. *J. Anim. Ecol. 38* : 341-60

_____ 1969b. A population study of house mice permanently inhabiting a reed bed in South Australia. *J. Anim. Ecol. 38* : 361-78

Nichols, J.D., Hestbeck, J.B. and Conley, W. 1979. Mathematical models and population cycles: a critical evaluation of a recent modelling effort. *J. Math. Biol. 8* : 259-63

Nikitina, N.A., Karulin, B.E. and Zen'kovich, N.S. 1972. The activity and territory of the common vole (*Microtus arvalis* Pall.). *Byull. Mosk. o-va. Ispyt. Prir. Otd. Biol. 77* : 55-64 [Not seen; cited by Madison 1984]

Nisbet, R.M. and Gurney, W.S.C. 1982. *Modelling Fluctuating Populations.* Wiley, Chichester, Sussex

van Noordwijk, A.R. 1984. Problems in the analysis of dispersal and a critique on its 'heritability' in the great tit. *J. Anim. Ecol.* 53 : 533-44

―――, van Balen, J.H. and Scharloo, W. 1981. Genetic and environmental variation in clutch size of the great tit. *Neth. J. Zool.* 31 : 342-72

Nunney, L. 1985a. Female-biased sex ratios : individual or group selection? *Evolution* 39 : 349-61

――― 1985b. Short time delays in population models: a role in enhancing stability. *Ecology* 66 : 1849-58

Nygren, J. 1980a. Allozyme variation in natural populations of field vole (*Microtus agrestis* L.). II. Survey of an isolated island population. *Hereditas* 93 : 107-14

――― 1980b. Allozyme variation in natural populations of field vole (*Microtus agrestis* L.). III. Survey of a cyclically varying population. *Hereditas* 93 : 125-36

Oakeshott, J.G., Chambers, G.K., Gibson, J.B. and Willcocks, D.A. 1981. Latitudinal relationships of esterase-6 and phosphoglucomutase gene frequencies in *Drosophila melanogaster*. *Heredity* 47 : 385-96

―――, Gibson, J.B., Anderson, P.R., Knibb, W.R., Anderson, D.G. and Chambers, G.K. 1982. Alcohol dehydrogenase and glycerol-3-phosphate dehydrogenase clines in *Drosophila melanogaster* on different continents. *Evolution* 36 : 86-96

Ohno, S., Stenius, C. and Christian, L. 1966. The XO as a normal female of the creeping vole (*Microtus oregoni*). *Chrom. Today* 1 : 182-7

Oksanen, L. 1983. Trophic exploitation and arctic phytomass patterns. *Amer. Natur.* 122 : 45-52

Okubo, A. 1980. *Diffusion and Ecological Problems: Mathematical Models.* Springer-Verlag, New York

Oliveras, D. and Novak, M. 1986. A comparison of paternal behaviour in the meadow vole *Microtus pennsylvanicus*, the pine vole *M. pinetorum*, and the prairie vole *M. ochrogaster*. *Anim. Behav.* 34 : 519-26

Orians, G.H. 1969. On the evolution of mating systems in birds and mammals. *Amer. Natur.* 103 : 589-603

Oring, L.W. 1982. Avian mating systems. *Avian Biology* 6 : 1-92

Orzack, S.H. 1985. Population dynamics in variable environments. V. The genetics of homeostasis revisited. *Amer. Natur.* 125 : 550-72

Oster, G.F. and Takashi, Y. 1974. Models for age specific interactions in a periodic environment. *Ecol. Monogr.* 44 : 483-501

――― and Wilson, E.O. 1978. *Caste and Ecology in the Social Insects.* Princeton University Press, Princeton

Ostfeld, R.S. 1985a. Limiting resources and territoriality in microtine rodents. *Amer. Natur.* 126 : 1-15

――― 1985b. Experimental analysis of aggression and spacing behavior in California voles. *Can. J. Zool.* 63 : 2277-82

――― 1986. Territoriality and mating system of California voles. *J. Anim. Ecol.* 55 : 691-706

――― 1987. On the distinction between female defense and resource defense polygyny. *Oikos* in press

――― and Klosterman, L.L. 1986. Demographic substructure in a California vole population inhabiting a patchy environment. *J. Mamm.* 67 : 693-704

―――, Lidicker, W.Z. and Heske, E.J. 1985. The relationship between habitat heterogeneity, space use, and demography in a population of California voles. *Oikos* 45 : 433-42

Ouellette, D.E. and Helsinger, J.F. 1980. Reingestion of feces by *Microtus pennsylvanicus. J. Mamm. 61* : 366-8

Owen, D.F. 1977. Latitudinal gradients in clutch size: an extension of David Lack's theory. In B. Stonehouse and C.M. Perrins (eds.) *Evolutionary Ecology*. University Park Press, Baltimore, pp. 171-9

_____ 1980. How plants may benefit from the animals that eat them. *Oikos 35* : 230-5

Packer, C. 1977. Reciprocal altruism in *Papio anubis*. *Nature 265* : 441-3

_____ 1985. Dispersal and inbreeding avoidance. *Anim. Behav. 33* : 676-8

_____ and Pusey, A.E. 1982. Cooperation and competition within coalitions of male lions: kin selection or game theory? *Nature 296* : 740-2

Page, R.E. and Bergerud, A.T. 1984. A genetic explanation for ten-year cycles of grouse. *Oecologia (Berl.) 64* : 54-60

Parker, G.A. 1970a. The reproductive behaviour and nature of sexual selection in *Scatophaga stercoraria* L. (Diptera: Scatophagidae). II. The fertilization rate and the spatial and temporal relationships of each sex around the site of mating and oviposition. *J. Anim. Ecol. 39* : 205-28

_____ 1970b. The reproductive behaviour and nature of sexual selection in *Scatophaga stercoraria* L. (Diptera: Scatophagidae). IX. Spatial distribution of fertilisation rates and evolution of male search strategy within the reproductive area. *Evolution 28* : 93-108

_____ 1978. Evolution of competitive mate searching. *Ann. Rev. Entomol. 23* : 173-96

_____ 1984. Evolutionarily stable strategies. In J.R. Krebs and N.B. Davies (eds.) *Behavioural Ecology : an Evolutionary Approach*. 2nd edn. Blackwell Scientific Publications, Oxford, pp. 30-61

_____ 1985. Population consequences of evolutionarily stable strategies. In R.M. Sibly and R.H. Smith (eds.) *Behavioural Ecology: Ecological Consequences of Adaptive Behaviour*. Blackwell Scientific Publications, Oxford, pp. 33-58

_____ and Rubenstein, D.I. 1981. Role assessment reserve strategy, and acquisition of information in asymmetric animal contests. *Anim. Behav. 29* : 221-40

_____ and Sutherland, W.J. 1986. Ideal free distributions when individuals differ in competitive ability : phenotype-limited ideal free models. *Anim. Behav. 34* : 1222-42

Parry, G.D. 1981. The meanings of r- and K- selection. *Oecologia (Berl). 48* : 260-4

Partridge, L. 1983. Non-random mating and offspring fitness. In P. Bateson (ed.) *Mate Choice*. Cambridge University Press, Cambridge, pp. 227-53

Paul, J.R. 1970. Observations on the ecology, populations, and reproductive biology of the pine vole, *Microtus pinetorum*, in North Carolina. *Illinois State Mus. Rep. Invest. 20* : 1-28

Pavone, L.V. and Boonstra, R. 1984. A technique for the surgical removal of a kidney from individuals of a feral population of small rodents. *Can. J. Zool. 62* : 2149-56

Pearson, O.P. 1971. Additional measurements of the impact of carnivores on California voles (*Microtus californicus*). *J. Mamm 52* : 41-9

_____ 1985. Predation. In R.H. Tamarin (ed.) *Biology of New World* Microtus. Special Publication No. 8 , American Society of Mammalogists, pp. 535-66

Pease, J.L., Vowles, R.H. and Keith, L.B. 1979. Interaction of snowshoe hares and woody vegetation. *J. Wildl. Manage. 43* : 43-60

Pedersen, H.C. 1984. Territory size, mating status, and individual survival of males in a fluctuating population of willow ptarmigan. *Ornis Scand. 15* : 197-203

Pehrson, Å. 1981. Winter food consumption and digestibility in caged mountain hares. In K. Myers and C.D. MacInnes (eds.) *Proceedings of the World Lagomorph Conference*. University of Guelph, Guelph, pp. 732-42

_____ and Lindlöf, B. 1984. Impact of winter nutrition on reproduction in captive mountain hares (*Lepus timidus*) (Mammalia: Lagomorpha). *J. Zool. Lond.* 204 : 201-9

Pelikán, J. 1982. Microtus arvalis on mown and unmown meadow. *Acta Sc. Nat. Brno,* 16(11): 1-36

Pendergast, B.A. and Boag, D.A. 1973. Seasonal changes in the internal anatomy of spruce grouse in Alberta. *Auk* 90 : 307-17

Perrins, C.M. 1965. Population fluctuations and clutch size in the great tit, *Parus major* L. *J. Anim. Ecol.* 34 : 601-47

_____ and Jones, P.J. 1974. The inheritance of clutch size in the great tit. *Condor* 76 : 225-9

Peters, R.H. 1983. *The Ecological Implications of Body Size.* Cambridge University Press, Cambridge.

Peterson, R.O., Page, R.E. and Dodge, K.M. 1984. Wolves, moose, and the allometry of population cycles. *Science* 224 : 1350-2

Petterborg, L.J. 1978. Effect of photoperiod on body weight in the vole, *Microtus montanus.* *Can. J. Zool.* 56 : 431-5

Pianka, E.R. 1970. On r- and K- selection. *Amer. Natur.* 104 : 592-7

Pickering, G. 1961. Language: the lost tool of learning in medicine and science. *The Lancet, 15 July 1961* : 115-9

Pickering, J., Getz, L.L. and Whitt, G.S. 1974. An esterase phenotype correlated with dispersal in *Microtus. Trans. Ill. Acad. Sci.* 67 : 471-5

Pinkel, D., Gledhill, B.L., Lake, S., Stephenson, D. and Van Dilla, M.A. 1982. Sex preselection in mammals? Separation of sperm bearing 'Y' and 'O' chromosomes in the vole *Microtus oregoni. Science* 218 : 904-6

Pinter, A.J. 1968. Effects of diet and light on growth, maturation and adrenal size in Microtus montanus. *Amer. J. Physiol.* 215 : 461-6

Pitelka, F.A. 1964. The nutrient recovery hypothesis for arctic microtine cycles. I. Introduction. In D.Crisp (ed.) *Grazing in Terrestrial and Marine Environments.* Blackwell Scientific Publications, Oxford, pp. 55-6

Pokki, J. 1981. Distribution, demography and dispersal of the field vole, *Microtus agrestis* (L.), in the Tvarminne archipelago, Finland. *Acta Zool. Fennica* 164 : 1-48

Porter, J.H. and Dueser, R.D. 1986. A test for suppression of body growth and sexual maturation in small male meadow voles (*Microtus pennsylvanicus*) in field enclosures. *Amer. Midl. Natur.* 115 : 181-90

Power, M.E. 1984. Habitat quality and the distribution of algae-grazing catfish in a Panamanian stream. *J. Anim. Ecol.* 53 : 357-74

Price, M.V. and Waser, N.M. 1979. Pollen dispersal and optimal out-crossing in *Delphinium nelsoni. Nature* 280 : 389-91

Price, T.D. and Grant, P.R. 1984. Life history traits and natural selection for small body size in a population of Darwin's finches. *Evolution* 38 : 483-94

_____ , _____ , Gibbs, H.L. and Boag, P.T. 1984. Recurrent patterns of natural selection in a population of Darwin's finches. *Nature* 309 : 787-9

Prout, T. 1981. A note on the island model with sex dependent migration. *Theoret. Appl. Genet.* 59 : 327-32

Pulliainen, E. 1982. Habitat selection and fluctuations in numbers in a population of the arctic hare (*Lepus timidus*) on a subarctic fell in Finnish Forest Lapland. *Z. Säugetierkunde* 47 : 168-74

_____ 1983. The refuge theory and habitat selection in the mountain hare on a subarctic fell in Finnish Forest Lapland. *Finnish Game Res.* 41 : 39-44

_____ and Keränen, J. 1979. Composition and function of beard lichen stores accumulated by bank voles, *Clethrionomys glareolus* Schreb. *Aquilo Ser. Zool.* 19 : 73-6

REFERENCES

———— and Tunkkari, P. 1983. Seasonal changes in the gut length of the willow grouse (*Lagopus lagopus*) in Finnish Lapland. *Annales Zoologica Fennici* 20 : 53-6

Pulliam, H.R. 1975. Diet optimization with nutrient constraints. *Amer. Natur.* 109 : 765-8

———— and Caraco, T. 1984. Living in groups: is there an optimal size? In J.R. Krebs and N.B. Davies (eds.) *Behavioural Ecology: an Evolutionary Approach.* 2nd edn, Blackwell Scientific Publications, Oxford, pp. 122-147

Pyke, G.H. 1979. The economics of territory size and time budget in the golden-winged sunbird. *Amer. Natur. 114* : 131-45

Quay, W.B. 1980. Greater pineal volume and latitudinal and chronobiological characteristics: exponential relationship and its biological interpretation. *Gen. Comp. Endocrinol. 41* : 340-8

———— 1984. Winter tissue changes and regulatory mechanisms in nonhibernating small mammals : a survey and evaluation of adaptive and non-adaptive features. In J.F. Merritt (ed.) *Winter Ecology of Small Mammals.* Special Publication No. 10, Carnegie Museum of Natural History, Pittsburgh, pp. 149-63

Randall, J.A. 1978. Behavioral mechanisms of habitat segregation between sympatric species of *Microtus* : habitat preference and interspecific dominance. *Behav. Ecol. Sociobiol. 3* : 187-202

Randolph, P.A., Randolph, J.C., Mattingly, K. and Foster, M.M. 1977. Energy costs of reproduction in the cotton rat, *Sigmodon hispidus.* *Ecology* 58 : 31-45

Rasa, O.A.E. and van den Höövel, H. 1984. Social stress in the field vole: differential causes of death in relation to behaviour and social structure. *Z. Tierpsychol. 65*: 108-33

Rasmuson, B., Rasmuson, M and Nygren, J. 1977. Genetically controlled differences in behaviour between cycling and non-cycling populations of field vole (*Microtus agrestis*). *Hereditas 87* : 33-42

Rausher, M.D. 1984. The evolution of habitat preferences in subdivided populations. *Evolution 38* : 598-608

Real, L.A. 1980. Fitness, uncertainty, and the role of diversification in evolution and behavior. *Amer. Natur. 115* : 623-38

Redfield, J.A., Taitt, M.J. and Krebs, C.J. 1978a. Experimental alteration of sex ratios in populations of *Microtus townsendii*, a field vole. *Can. J. Zool. 56* : 17-27

————, ———— and ———— 1978b. Experimental alteration of sex ratios in populations of *Microtus oregoni*, the creeping vole. *J. Anim. Ecol. 47* : 55-69

Reich, L.M. and Tamarin, R.H. 1980. Trap use as an indicator of social behavior in mainland and island voles. *Acta theriol. 25* : 295-307

————, Wood, K.M., Rothstein, B.E. and Tamarin, R.H. 1982. Aggressive behaviour of male *Microtus breweri* and its demographic implications. *Anim. Behav. 30* : 117-22

Reichert, S.E. 1981. The consequences of being territorial : spiders, a case study. *Amer. Natur. 117*: 871-92

Reznick, D. 1985. Costs of reproduction: an evaluation of the empirical evidence. *Oikos* 44 : 257-67

Rhoades, D.F. and Cates, R.G. 1976. Towards a general theory of plant antiherbivore chemistry. *Recent Adv. Phytochem. 10* : 168-213

Richmond, M.E. and Conaway, C.H. 1969. Induced ovulation and oestrus in *Microtus ochrogaster. J. Reprod. Fertil. Suppl. 6* : 357-76

———— and Stehn, R. 1976. Olfaction and reproductive behavior in Microtine rodents. In R.L. Doty (ed.) *Mammalian Olfaction, Reproductive Processess and Behavior.* Academic Press, New York, pp. 197-217

Richter, W. 1984. Nestling survival and growth in the yellow-headed blackbird, *Xanthocephalus xanthocephalus. Ecology 65* : 597-608

Ricklefs, R.E. 1977. On the evolution of reproductive strategies in birds: reproductive effort. *Amer. Natur. 111* : 453-78
_____ 1980. Geographical variation in clutch size among passerine birds: Ashmole's hypothesis. *Auk 97* : 38-49
Ridley, M. 1983. *The Explanation of Organic Diversity : The Comparative Method and Adaptations for Mating*. Clarendon Press, Oxford
Riggs, L.A. 1979. Experimental studies of dispersal in the California vole, *Microtus californicus*. Ph.D. thesis, University of California at Berkeley
Rissman, E.F., Sheffield, S.D., Kretzmann, M.B., Fortune, J.E. and Johnston, R.E. 1984. Chemical cues from families delay puberty in male California voles. *Biol. Reprod. 31* : 324-31
Rivers, J.P.W. and Crawford, M.A. 1974. Maternal nutrition and sex ratio at birth. *Nature* 252 : 297-8
Robertson, O.H. 1961. Prolongation of the life of Kokanee salmon (*Oncoryhnchus nerka kennerlyi*) by castration before beginning of gonadal development. *Proc. natn. Acad. Sci. USA 47* : 609-21
_____ and Wexler, B.C. 1957. Pituitary degeneration and adrenal tissue hyperplasia in spawning Pacific salmon. *Science 125* : 1295-6
_____ and _____ 1960. Histological changes in the organs and tissues of migrating and spawning Pacific salmon (Genus *Oncorhynchus*). *Endocrinology 66* : 222-39
Rodd, F.H. and Boonstra, R. 1984. The spring decline in the meadow vole, *Microtus pennsylvanicus*: the effect of density. *Can. J. Zool. 62* : 1464-73
Rose, M.R. 1983. Theories of life-history evolution. *Amer. Zool. 23* : 15-23
_____ 1984a. The evolution of animal senescence. *Can. J. Zool. 62* : 1661-7
_____ 1984b. Genetic covariation in *Drosophila* life history: untangling the data. *Amer. Natur. 123* : 565-9
_____ 1985. Life history evolution with antagonistic pleiotropy and overlapping generations. *Theoret. Popul. Biol. 28* : 342-58
_____ and Charlesworth, B. 1980. A test of evolutionary theories of senescence. *Nature* 287 : 141-2
_____ and _____ 1981a. Genetics of life history in *Drosophila melanogaster*. I. Sib analysis of adult females. *Genetics 97* : 173-86
_____ and _____ 1981b. Genetics of life history in *Drosophila melanogaster*. II. Exploratory selection experiments. *Genetics 97* : 187-96
Rose, R.K. and Birney, E.C. 1985. Community ecology. In R.H. Tamarin (ed.), *Biology of New World* Microtus. Special Publication No. 8 , American Society of Mammalogists, pp. 310-39
_____ and Dueser, R.D. 1980. Lifespan of Virginia meadow voles. *J. Mamm. 61* : 760-3
_____ and Gaines, M.S. 1978. The reproductive cycle of *Microtus ochrogaster* in eastern Kansas. *Ecol. Monogr. 48* : 21-42
Rosenzweig, M.L. and Abramsky, Z. 1980. Microtine cycles: the role of habitat heterogeneity. *Oikos 34* : 141-6
Rothstein, B.E. and Tamarin, R.H. 1977. Feeding behavior of the insular beach vole, *Microtus breweri*. *J. Mamm. 58* : 84-5
Roughgarden, J.D. 1977. Patchiness in the spatial distribution of a population caused by stochastic fluctuations in resources. *Oikos 29* : 52-9
Rubenstein, D.I. 1982. Risk, uncertainty and evolutionary strategies. In Kings' College Sociobiology Group (eds.) *Current Problems in Sociobiology*. Cambridge University Press, Cambridge, pp. 91-111
Saitoh, T. 1981. Control of female maturation in high density populations of the red-backed vole, *Clethrionomys rufocanus bedfordiae*. *J. Anim. Ecol. 50* : 79-87

REFERENCES

_____ 1983. Survival rate and mobility in an enclosed population of red-backed vole, *Clethrionomys rufocanus bedfordiae*. *Acta theriol. 19* : 301-15
Sanders, E.H., Gardner, P.D., Berger, P.J. and Negus, N.C. 1981. 6-methoxybenzoxazolinone : a plant derivative that stimulates reproduction in *Microtus montanus*. *Science 214* : 67-9
Savory, C.J. 1983. Selection of heather age and chemical composition by red grouse in relation to physiological state, season and time of day. *Ornis Scandinavica 14* : 135-43
Schadler, M.H. 1981. Postimplantation abortion in pine voles (*Microtus pinetorum*) induced by strange males and pheromones of strange males. *Biol. Reprod. 25* : 295-7
_____ 1983. Male siblings inhibit reproductive activity in female pine voles, *Microtus pinetorum*. *Biol. Reprod. 28* : 1137-9
_____ 1985. Strange males cause death or suppression of growth in infant pine voles, *Microtus pinetorum*. *J. Mamm. 66* : 387-90
Schaffer, W.M. 1974a. Selection for optimal life histories: the effect of age structure. *Ecology 55* : 291-303
_____ 1974b. Optimal reproductive effort in fluctuating environments. *Amer. Natur. 108* : 783-90
_____ 1981. On reproductive value and fitness. *Ecology 62* : 1683-5
_____ 1984. Stretching and folding in lynx fur returns: evidence for a strange attractor in nature? *Amer. Natur. 124* : 798-820
_____ 1985. Order and chaos in ecological systems. *Ecology 66* : 93-106
_____ and Kot, M. 1985. Do strange attractors govern ecological systems? *BioScience 35* : 342-50
_____ and _____ 1986. Chaos in ecological systems: the coals that Newcastle forgot. *TREE 1* : 58-63
_____ and Tamarin, R.H. 1973. Changing reproductive rates and population cycles in lemmings and voles. *Evolution 27* : 111-24
Schierwater, B. and Klingel, H. 1986. Energy costs of reproduction in the djungarian hamster *Phodopus sungorus* under laboratory and seminatural conditions. *Oecologia (Berl.) 69* : 144-7
Schoener, T.W. 1983. Simple models of optimal feeding-territory size: a reconciliation. *Amer. Natur. 121* : 608-29
Schroeder, M.A. 1986. The fall phase of dispersal in juvenile spruce grouse. *Can. J. Zool. 64* : 16-20
Schultz, A.M. 1964. The nutrient recovery hypothesis for arctic microtine cycles. II. Ecosystem variables in relation to arctic microtine cycles. In D.Crisp (ed.) *Grazing in Terrestrial and Marine Environments*. Blackwell Scientific Publications, Oxford, pp. 57-68
Schwagmeyer, P.L. 1979. The Bruce effect: an evaluation of male/female advantages. *Amer. Natur. 114* : 932-8
Scott, M.P and Tan, T.N. 1985. A radiotracer technique for the determination of male mating success in natural populations. *Behav. Ecol. Sociobiol. 17* : 29-33
Scotts, D.J. 1983. The social organization of *Antechinus stuartii* (Macleay)(Marsupialia, Dasyuridae) at Sherbrooke Forest, Victoria. B.Sc. (Hons) thesis, Monash University, Victoria, Australia
Seabloom, R.W. 1985. Endocrinology. In R.H. Tamarin (ed.) *Biology of New World Microtus*. Special Publication No. 8, American Society of Mammalogists, pp. 685-724
Seger, J. 1983. Partial bivoltinism may cause alternating sex ratio biases that favour eusociality. *Nature 301* : 59-62

_____ 1985. Unifying genetic models for the evolution of female choice. *Evolution* 39 : 1185-93

_____ and Trivers, J. 1985. Asymmetry in the evolution of female mating preferences. *Nature 319* : 771-3

Selye, H. 1936. A syndrome produced by diverse nocuous agents. *Nature 138* : 32

Selye, H. 1946. The general-adaptation-syndrome and diseases of adaptation. *J. Clin. Endocr. Metab. 6* : 117-230

Semb-Johansson, A., Wiger, R. and Engh, C.E. 1979. Dynamics of freely growing, confined populations of the Norwegian lemming *Lemmus lemmus*. *Oikos 33* : 246-60

Service, P.M. and Rose, M.R. 1985. Genetic covariation among life-history components: the effect of novel environments. *Evolution 39* : 943-5

Shapiro, L.E., Austin, D., Ward, S.E. and Dewsbury, D.A. 1986. Familiarity and female mate choice in two species of voles (*Microtus ochrogaster* and *Microtus montanus*). *Anim. Behav. 34* : 90-7

_____ and Dewsbury, D.A. 1986. Male dominance, female choice and male copulatory behavior in two species of voles (*Microtus ochrogaster* and *Microtus montanus*). *Behav. Ecol. Sociobiol. 18* : 267-74

Sheridan, M. and Tamarin, R.H. 1985. Genetic variation of salivary amylase in meadow voles. *J. Mamm 66* : 821-3

Sherman, P.W. 1981. Reproductive competition and infanticide in Belding's ground squirrels and other animals. In R.D. Alexander and D.W. Tinkle (eds.) *Natural Selection and Social Behavior*. Chiron Press, New York, pp. 311-31

Shields, W.M. 1983. Optimal inbreeding and the evolution of philopatry. In I.R. Swingland and P.J. Greenwood (eds.) *The Ecology of Animal Movement*. Clarendon Press, Oxford, pp. 132-59

Shvarts, S.S., Pokrovski, A.V., Istchenko, V.G., Olenjec, V.G., Ovtschinnikova, N.A. and Pjastoloba, O.A. 1964. Biological peculiarities of seasonal generations of rodents, with special reference to the problem of senescence in mammals. *Acta Theriol. 8* : 11-43

Sibly, R. and Calow, P. 1984. Direct and absorption costing in the evolution of life cycles. *J. theor. Biol. 111* : 463-73

_____ and _____ 1985. Classification of habitats by selection pressures: a synthesis of life-cycle and r/K theory. In R.M. Sibly and R.H. Smith (eds.) *Behavioural Ecology: Ecological Consequences of Adaptive Behaviour*. Blackwell Scientific Publications, Oxford, pp. 75-90

Silk, J.B. 1984. Local resource competition and the evolution of male-biased sex ratios. *J. theor. Biol. 108* : 203-13

Sinclair, A.R.E., Krebs, C.J. and Smith, J.N.M. 1982. Diet quality and food limitation in herbivores: the case of the snowshoe hare. *Can. J. Zool. 60* : 889-897

Singleton, G.R. and Hay, D.A. 1982. A genetic study of male social aggression in wild and laboratory mice. *Behavior Genetics 12* : 435-48

_____ and _____ 1983. The effect of social organization on reproductive success and gene flow in colonies of wild house mice, *Mus musculus*. *Behav. Ecol. Sociobiol. 12* : 49-56

Slatkin, M. 1985. Gene flow in natural populations. *Ann. Rev. Ecol. Syst. 16* : 393-430

Smith, C.H. and Davis, J.M. 1981. A spatial analysis of wildlife's ten year cycle. *J. Biogeog. 8* : 27-35

Smith, J.N.M. 1981. Does high fecundity reduce survival in song sparrows. *Evolution 35* : 1142-8

Smith, M.H., Garten, C.T. and Ramsey, P.R. 1975. Genic heterozygosity and population dynamics in small mammals. In C.L. Markert (ed.) *Isozymes. IV. Genetics and Evolution.* Academic Press, New York, pp. 85-102

———, Manlove, M.N. and Joule, J. 1978. Spatial and temporal dynamics of the genetic organisation of small mammal populations. In D.P. Snyder (ed.) *Populations of Small Mammals under Natural Conditions.* Pyamatuning Ecology Lab, Pittsburgh, pp. 99-113

Smith, R.H. and Sibly, R. 1985. Behavioural ecology and population dynamics : towards a synthesis. In R.M. Sibly and R.H. Smith (eds.) *Behavioural Ecology: Ecological Consequences of Adaptive Behaviour.* Blackwell Scientific Publications, Oxford, pp. 577-91

Smith-Gill, S.J. 1983. Developmental plasticity: developmental conversion versus phenotypic modulation. *Amer. Zool.* 23 : 47-55

Smolen, M.J. and Keller, B.L. 1979. Survival, growth and reproduction of progeny isolated from low and high density populations of *Microtus montanus. J. Mamm.* 60 : 265-79

Snipes, R.L. 1979. Anatomy of the cecum of the vole, *Microtus agrestis. Anat. Embryol. 157* : 181-203

Southwood, T.R.E. 1977. Habitat, the templet for ecological strategies? *J. Anim. Ecol.* 46 : 337-65

———, Brown, V.K. and Reader, P.M. 1983. Continuity of vegetation in space and time : a comparison of insects' habitat templet in different successional stages. *Res. Popul. Ecol. Suppl. 3* : 61-74

Sperber, I., Björnhag, G. and Ridderstråle, Y. 1983. Function of proximal colon in lemming and rat. *Swedish J. agric. Res. 13* : 243-56

Stearns, S.C. 1976. Life history tactics: a review of the ideas. *Q. Rev. Biol. 51* : 3-47

——— 1977. The evolution of life history traits : a critique of the theory and a review of the data. *Ann. Rev. Ecol. Syst. 8* : 145-71

——— 1983. Introduction to the symposium: the interface of life-history evolution, whole-organism ontogeny and quantitative genetics. *Amer. Zool.* 23 : 3-4

——— and Crandall, R.E. 1981a. Quantitative predictions of delayed maturity. *Evolution* 35 : 455-63

——— and ——— 1981b. Bet-hedging and persistence as adaptations of colonizers. In G.G.E. Scudder and J.L. Reveal (eds) *Evolution Today.* Hunt Institute for Botanical Documentation, Pittsburgh, pp. 371-83

Stehn, R.A. and Jannett, F.J. 1981. Male-induced abortion in various microtine rodents. *J. Mamm.* 62 : 369-72

——— and Richmond, M.E. 1975. Male-induced pregnancy termination in the prairie vole, *Microtus ochrogaster. Science 187* : 1211-13

Stenseth, N.C. (ed.) 1977. Population Dynamics of *Microtus agrestis. Oikos* 29 : 445-641

——— 1978a. Demographic strategies in fluctuating populations of small rodents. *Oecologia (Berl.) 33* : 149-72

——— 1978b. Is the female biased sex ratio in wood lemming *Myopus schisticolor* maintained by cyclic inbreeding. *Oikos 30* : 83-9

——— 1978c. Energy balance and the Malthusian parameter, m, of grazing small rodents. *Oecologia (Berl.) 32* : 37-55

——— 1980. Spatial heterogeneity and population stability: some evolutionary consequences. *Oikos 35* : 165-84

——— 1981. On Chitty's theory for fluctuating populations : the importance of genetic polymorphism in the generation of regular density cycles. *J. theor. Biol. 90* : 9-36

——— 1983. Causes and consequences of dispersal in small mammals. In I.R. Swingland and P.J. Greenwood (eds.) *The Ecology of Animal Movement.* Clarendon Press,

Oxford, pp. 63-101

———— 1985. Professor Olavi Kalela. *Ann. Zool. Fennici* 22 : 208-10

———— 1986. On the interaction between stabilizing social and destabilizing trophic factors in small rodent populations. *Theor. Popul. Biol.* 29 : 365-84

———— and Framstad, E. 1980. Reproductive effort and optimal reproductive rates in small rodents. *Oikos* 34 : 23-34

————, ————, Migula, P., Trojan, P. and Wojciechowska-Trojan, B. 1980. Energy models for the common vole *Microtus arvalis*: energy as a limiting resource for reproductive output. *Oikos* 34 : 1-22

————, Gustafsson, T.O., Hansson, L. and Ugland, K.I. 1985. On the evolution of reproductive rates in microtine rodents. *Ecology* 66 : 1795-1808

———— and Lidicker, W.Z. 1987. Presaturation and saturation dispersal ten years later: some theoretical considerations. In N.C. Stenseth and W.Z. Lidicker (eds.) *Animal Dispersal : Small Mammals as a Model*. Chapman and Hall, London, pp. 000-000

———— and Ugland, K.I. 1985. On the evolution of demographic strategies in populations with equilibrium and cyclic densities. *Math. Biosc.* 74 : 89-109

Stoddart, D.M. 1970. Individual range, dispersion and dispersal in a population of water voles (*Microtus agrestis*). *J. Anim. Ecol.* 39 : 403-25

Sullivan, T.P. 1980. Comparative demography of *Peromyscus maniculatus* and *Microtus oregoni* populations after logging and burning of coastal forest habitats. *Can. J. Zool.* 58 : 2252-9

———— and Krebs, C.J. 1981. *Microtus* population biology: demography of *M. oregoni* in southwestern British Columbia. *Can. J. Zool.* 59 : 2092-2102

Sutherland, W.J. and Parker, G.A. 1985. Distribution of unequal competitors. In R.M. Sibly and R.H. Smith (eds.) *Behavioural Ecology: Ecological Consequences of Adaptive Behaviour*. Blackwell Scientific Publications, Oxford, pp. 255-73

Tähkä, K.M., Roukonen, A., Wallgren, H. and Teräväinen, T. 1983. Temporal changes in testicular steroidogenesis in juvenile bank voles (*Clethrionomys glareolus* Schreber) subjected to different photoperiods. *Endocrinology* 112 : 1420-6

Taitt, M.J. 1981. The effect of extra food on small rodent populations. I Deermice (*Peromyscus leucopus*). *J. Anim. Ecol.* 50 : 111-24

———— 1985a. Experimental analysis of spacing behaviour in the vole, *Microtus townsendii*. In R.M. Sibly and R.H. Smith (eds.) *Behavioural Ecology: Ecological Consequences of Adaptive Behaviour*. Blackwell Scientific Publications, Oxford, pp. 313-7

———— 1985b. An experimental test of simultaneous settlement in the vole, *Microtus townsendii*. *Can. J. Zool.* 63 : 1663-7

———— and Krebs, C.J. 1981. The effect of extra food on small rodent populations. II. Voles (*Microtus townsendii*). *J. Anim. Ecol.* 50 : 125-37

———— and ———— 1983. Predation, cover and food manipulations during a spring decline of *Microtus townsendii*). *J. Anim. Ecol.* 52 : 837-48

———— and ———— 1985. Population dynamics and cycling. In R.H. Tamarin (ed.) *Biology of New World* Microtus. Special Publication No. 8, American Society of Mammalogists, pp. 567-620

Tamarin, R.H. 1977a. Dispersal in island and mainland voles. *Ecology* 58 : 1044-54

———— 1977b. Demography of the beach vole (*Microtus breweri*) and the meadow vole (*Microtus pennsylvanicus*) on southeastern Massachusetts. *Ecology* 58 : 1310-21

———— 1978a. Dispersal, population regulation, and K-selection in field mice. *Amer. Natur.* 112 : 545-55

———— 1978b. A defense of single-factor models of population regulation. In D.P Snyder (ed.) *Populations of Small Mammals under Natural Conditions*. Pymatuning Laboratory of Ecology, Pittsburgh, pp. 159-62

_____ 1980. Dispersal and population regulation in rodents. In M.N. Cohen, R.S. Malpass and H.G. Klein (eds.) *Biosocial Mechanisms of Population Regulation.* Yale University Press, New Haven, pp. 117-33

_____ 1981. Hip glands in wild-caught *Microtus pennsylvanicus. J. Mamm. 62* : 421

_____ 1983. Animal population regulation through behavioral interactions. In J.F. Eisenberg and D.V. Kleiman (eds.) *Advances in the Study of Animal Behavior.* American Society of Mammalogists, Special Publication No. 7, pp. 698-720

_____ 1984. Body mass as a criterion of dispersal in voles: a critique. *J. Mamm. 65* : 691-2

_____ (ed.)1985. *Biology of the New World* Microtus. Special Publication No. 8, American Society of Mammalogists

_____ and Kunz, T. 1974. *Microtus breweri. Mammalian Species 45* : 1-3

_____, Reich, L.M. and Moyer, C.A. 1984. Meadow vole cycles within fences. *Can. J. Zool. 62* : 1796-1804

_____, Sheridan, M. and Levy, C.K. 1983. Determining matrilineal kinship in natural populations of rodents using radionuclides. *Can. J. Zool. 61* : 271-4

Tanemura, M. and Hasegawa, M. 1980. Geometrical models of territory. I. Models for synchronous and asynchronous settlement of territories. *J. Theor. Biol. 82* : 477-96

Tanner, J.T. 1975. The stability and the intrinsic growth rates of prey and predator populations. *Ecology 56* : 855-67

Tast, J. 1966. The root vole, *Microtus oeconomus* (Pallas), as an inhabitant of seasonally flooded land. *Ann. Zool. Fennica 3* : 127-71

_____ 1974. The food and feeding habits of the root vole, *Microtus oeconomus*, in Finnish Lapland. *Aquilo Ser. Zool. 15* : 25-32

_____ 1980. Breeding season and litter size of the field vole *Microtus agrestis* at Kilpisjärvi, Finnish Lapland. *Kilpisjärvi Notes 4* : 8-11

_____ 1984. Winter success of root voles, *Microtus oeconomus*, in relation to population density and food conditions at Kilpisjärvi, Finnish Lapland. In J.F. Merritt (ed.). *Winter Ecology of Small Mammals.* Special Publication No. 10, Carnegie Museum of Natural History, Pittsburgh, pp. 59-66

_____ and Kalela, O. 1971. Comparisons between rodent cycles and plant production in Finnish Lapland. *Ann. Acad. Sci. Fennica (A IV) 186* : 1-14

Taylor, P.D. 1981. Intra-sex and inter-sex sibling interactions as sex ratio determinants. *Nature 291* : 64-6

Taylor, R.A.J. and Taylor, L.R. 1979. A behavioural model for the evolution of spatial dynamics. In R.M. Anderson, B.D. Turner and L.R. Taylor (eds.) *Population Dynamics.* Blackwell Scientific Publications, Oxford, pp. 1-27

Templeton, A.R. and Rothman, E.D. 1974. Evolution in heterogeneous environments. *Amer. Natur. 108* : 409-28

Thomas, J.A. and Birney, E.C. 1979. Parental care and mating system of the prairie vole, *Microtus ochrogaster. Behav. Ecol. Sociobiol 5* : 171-86

Tishkov, A.A., Cotfrid, A.B. and Sheftel, B.I. 1978. New data on the feeding habits of the wood lemming [in Russian]. *Byull. Mosk. Osshch. Ispyt. Prir., Otd. Biol. 83* : 22-6

Trivers, R.L. 1972. Parental investment and sexual selection. In B. Campbell (ed.) *Sexual Selection and the Descent of Man 1871-1971.* Heinemann, London, pp136-79

_____ 1974. Parent-offspring conflict. *Amer. Zool. 14* : 249-65

_____ 1985. *Social Evolution.* Benjamin/Cummings, Menlo Park, California

_____ and Willard, D.E. 1973. Natural selection of parental ability to vary the sex ratio of offspring. *Science 179* : 90-2

Turner, A.K. 1983. Time and energy constraints on the brood size of swallows, *Hirundo rustica*, and *Riparia riparia. Oecologia (Berl.) 59* : 331-8

Uchmanski, J. 1983. The effect of emigration on population stability: a generalization of the model of regulation of animal numbers, based on individual differences. *Oikos* 41 : 49-56

Ugland, K.I. and Stenseth, N.C. 1985. On the evolution of reproductive rates in populations with equilibrium and cyclic densities. *Math. Biosci.* 74 : 59-87

Ure, D.C. 1984. Autumn mass dynamics of red-backed voles (*Clethrionomys gapperi*) in Colorado in relation to photoperiod cues and temperature. In J.F. Merritt (ed.) *Winter Ecology of Small Mammals.* Special Publication No. 10, Carnegie Museum of Natural History, Pittsburgh, pp. 193-200

Valentine, G.L and Kirkpatrick, R.L. 1970. Seasonal changes in reproductive and related organs in the pine vole, *Microtus pinetorum*, in southwestern Virginia. *J. Mamm.* 51 : 553-60

Vance, R.R. 1984. The effect of dispersal on population stability in one-species, discrete-space population growth models. *Amer. Natur.* 123 : 230-54

Van den Assem, J. 1967. Territory in the three-spined stickleback *Gasterosteus aculeatus. Behaviour Suppl.* 16 : 1-164

Vandenbergh, J.G. 1969. Male odor accelerates female sexual maturation in mice. *Endocrinology* 84 : 658-60

────── 1971. The influence of the social environment on sexual maturation in male mice. *J. Reprod. Fertil.* 24 : 383-90

────── 1980. The influence of pheromones on puberty in rodents. In D. Muller-Schwarze and R.M. Silverstein (eds). *Chemical Signals.* Plenum Press, London, pp. 229-41

──────, Finlayson, J.S., Dobrogosz, W.J., Dills, S.S. and Kost, T.A. 1976. Chromatographic separation of puberty accelerating pheromone from male mouse urine. *Biol. Reprod.* 15 : 260-5

Van Horne, B. 1983. Density as a misleading indicator of habitat quality. *J. Wildl. Manage.* 47 : 893-901

Vaughan, M.R. and Keith, L.B. 1980. Breeding by juvenile snowshoe hares. *J. Wildl. Manage* 44 : 948-51

Vaughan, T.A. and Czaplewski, N.J. 1985. Reproduction in Stephens' woodrat : the wages of folivory. *J. Mamm.* 66 : 429-43

Vehrencamp, S.L. and Bradbury, J.W. 1984. Mating systems and ecology. In J.R. Krebs and N.B. Davies (eds.) *Behavioural Ecology : an Evolutionary Approach.* 2nd edn. Blackwell Scientific Publications, Oxford, pp. 251- 78

Verner, J. 1964. Evolution of polygamy in the long-billed marsh wren. *Evolution 18* : 252-61

────── and Willson, M.F. 1966. The influence of habitats on mating systems of North American passerine birds. *Ecology* 47 : 557-67

Verner, L. and Getz, L.L. 1985. Significance of dispersal in fluctuating populations of *Microtus ochrogaster* and *M. pennsylvanicus. J. Mamm.* 66 : 338-47

Vickery, W.L. 1979. Food consumption and preferences in wild populations of *Clethrionomys gapperi* and *Napaeozapus insignis. Can. J. Zool.* 57 : 1536-42

Viitala, J. 1977. Social organization in cyclic subarctic populations of the voles *Clethrionomys rufocanus* (Sund.) and *Microtus agrestis* (L.) *Ann. Zool. Fennici* 14 : 53-93

────── 1980. Myyrien sosiologiasta Kilpisjärviellä. *Luonnon Tutkija 84* : 31-4

────── 1984. The red vole, *Clethrionomys rutilus* (Pall.), as a subordinate member of the rodent community at Kilpisjärvi, Finnish Lapland. *Acta Zool. Fennica* 172 : 67-70

────── and Hoffmeyer, I. 1985. Social organization in *Clethrionomys* compared with *Microtus* and *Apodemus*: social odours, chemistry and biological effects. *Ann. Zool. Fennici* 22 : 359-71

Vorontsov, N.N. 1960. The ways of food specialization and evolution of the alimentary system in Muroidea. In J. Kratchovil and J. Pelikan (eds.) *Symposium Theriologicum: Proceedings of the International Symposium on Methods of Mammalogical Investigation*. Czechoslovak Academy of Sciences, Prague, pp. 360-77

Vrba, E.S. 1983. Macroevolutionary trends: new perspectives on the role of adaptation and incidental effect. *Science 221* : 387-9

Wahlund, S. 1928. Zuzammensetzung von Populationen und Korrelationserscheinungen vom Standpunkt der Verebungslehre aus betrachet. *Hereditas 11* : 65-106

Warkowska-Dratnal, H. and Stenseth, N.C. 1985. Dispersal and the microtine cycle: comparison of two hypotheses. *Oecologia (Berl.) 65* : 468-77

Waser, N.M. and Price, M.V. 1983. Optimal and actual outcrossing in plants, and the nature of plant-pollinator interactions. In C.E. Jones and R.J. Little (eds.) *Handbook of Experimental Pollination Biology*. Scientific and Academic Editions, New York, pp. 341-59

Waser, P.M. 1981. Sociality or territorial defense? The influence of resource renewal. *Behav. Ecol. Sociobiol. 8* : 231-7

Watson, A. 1967. Social status and population regulation in the red grouse (*Lagopus lagopus scoticus*). *Proc. R. Soc. Popul. Study Grp. 2* : 22-30

_____ and Jenkins, D. 1968. Experiments on population control by territorial behaviour in red grouse. *J. Anim. Ecol. 37* : 595-614

_____ and Moss, R. 1970. Dominance, spacing behaviour and aggression in relation to population limitation in vertebrates. In A. Watson (ed.) *Animal Populations in Relation to their Food Resources*. Blackwell Scientific Publications, Oxford, pp. 167-218

_____ and _____ 1980. Advances in our understanding of the population dynamics of red grouse from a recent fluctuation in numbers. *Ardea 68* : 103-11

_____ , _____ , Rothery, P. and Parr, R. 1984. Demographic causes and predictive models of population fluctuations in red grouse. *J. Anim. Ecol. 53* : 639-62

_____ , _____ and Parr, R. 1984. Effects of food enrichment on numbers and spacing behaviour of red grouse. *J. Anim. Ecol. 53* : 663-78

_____ , _____ , Phillips, J. and Parr, R. 1977. The effect of fertilizers on red grouse stocks on Scottish moors grazed by sheep, cattle and deer. In P. Pesson and M.G. Birkan (eds.) *Ecologie du Petit Gibier et Aménagement des Chasses*. Gauthier-Villars, Paris, pp. 193-212

_____ and O'Hare, P.J. 1979. Red grouse populations on experimentally treated and untreated Irish bog. *J. appl. Ecol. 16* : 433- 52

Watts, C.H.S. 1970. A field experiment on intraspecific interactions in the red-backed vole *Clethrionomys glareolus*. *J. Mamm. 51* : 341-7

Webster, A.B. and Brooks, R.J. 1981. Social behaviour of *Microtus pennsylvanicus* in relation to seasonal changes in demography. *J. Mamm. 62* : 738-51

_____ , Gartshore, R.G. and Brooks, R.J. 1981. Infanticide in the meadow vole, *Microtus pennsylvanicus*: significance in relation to social system and population cycling. *Behav. Neur. Biol. 31* : 342-7

Wecker, S.C. 1963. The role of early experience in habitat selection by the prairie deer mouse, *Peromyscus maniculatus bairdi*. *Ecol. Monogr. 33* : 306-25

Wellings, P.W., Leather, S.R. and Dixon, A.F.G. 1980. Seasonal variation in reproductive potential: a programmed feature of aphid life cycles. *J. Anim. Ecol. 49* : 975-86

Werren, J.H. 1983. Sex ratio evolution under local mate competition in a parasitic wasp. *Evolution 37* : 116-24

_____ and Charnov, E.L. 1978. Facultative sex ratios and population dynamics. *Nature 272* : 349-50

_____ and Taylor, P.D. 1984. The effects of population recruitment on sex ratio selection. *Amer. Natur. 124* : 143-8

West, S.D. 1977. Midwinter aggregations in northern red-backed voles (*Clethrionomys rutilus*). Can. J. Zool. 55 : 1404-9

_____ 1982. Dynamics of colonization and abundance in central Alaskan populations of the northern red-backed vole, *Clethrionomys rutilus*. *J. Mamm. 63* : 128-43

_____ and Dublin, H.T. 1984. Behavioral strategies of small mammals under winter conditions: solitary or social? In J.F. Merritt (ed.) *Winter Ecology of Small Mammals*. Special Publication No. 10, Carnegie Museum of Natural History, Pittsburgh, pp. 293-99

Westlin, L.M. 1982a. Sterile matings at the beginning of the breeding season in *Clethrionomys glareolus* and Microtus agrestis. *Can. J. Zool. 60* : 2568-70

_____ 1982b. Increased fertility in young primiparous female bank voles, *Clethrionomys glareolus*, treated with prolactin and progesterone after mating. *J. Reprod. Fertil. 66* : 113-5

_____ and Nyholm, N.E.I. 1982. Sterile matings initiate the breeding season in the bank vole, *Clethrionomys glareolus*. A field and laboratory study. *Can. J. Zool. 60* : 387-91

_____ and Gustafsson, T.O. 1984. Influence of age and artificial vaginal stimulation on fertility in female bank voles (*Clethrionomys glareolus*). *J. Reprod. Fertil. 71* : 103-6

Westoby, M. 1978. What are the biological bases of varied diets? *Amer. Natur. 112* : 627-31

Whitham, T.G. 1980. The theory of habitat selection: examined and extended using *Pemphigus* aphids. *Amer. Natur. 115* : 449-66

Widdowson, E.M. 1976. The response of the sexes to nutritional stress. *Proc. Nutr. Soc. 35* : 175-80

Wiens, J.A. 1976. Population responses to patchy environments. *Ann. Rev. Ecol. Syst. 7* : 81-120

Wiley, R.H. 1974. Evolution of social organization and life-history patterns among grouse. *Q. Rev. Biol. 49* : 201-27

_____ 1983. The evolution of communication : information and manipulation. In T.R. Halliday and P.J.B. Slater (eds.) *Communication: Animal Behaviour, Vol. 2*. Blackwell Scientific Publications, Oxford, pp. 156-89

Williams, G.C. 1966a. *Adaptation and Natural Selection*. Princeton University Press, Princeton.

_____ 1966b. Natural selection, the costs of reproduction, and a refinement of Lack's principle. *Amer. Natur. 100* : 687-90

_____ 1979. The question of adaptive sex ratio in outcrossed vertebrates. *Proc. R. Soc. Lond. B205* : 567-80

Williams, J. 1985. Statistical analysis of fluctuations in red grouse bag data. *Oecologia (Berl.) 65* : 269-72

Wilson, A.C., Cann, R.L., Carr, S.M., George, M., Gyllensten, U.B., Helm-Bychowski, K.M., Higuchi, R.G., Palumbi, S.R., Prager, E.M., Sage, R.D. and Stoneking, M. 1985. Mitochondrial DNA and two perspectives on evolutionary genetics. *Biol. J. Linn. Soc. 26* : 375-400

Wilson, D.S. 1980. *The Natural Selection of Populations and Communities*. Benjamin/Cummings, Menlo Park, California.

_____ 1983. The group selection controversy: history and current status. *Ann. Rev. Ecol. Syst. 14* : 159-87

_____ and Colwell, R.K. 1981. Evolution of the sex ratio in structured demes. *Evolution 35* : 882-97

REFERENCES

Wilson, E.O. 1971. *The Insect Societies.* Belknap Press, Cambridge, Massachussets
_____ 1975. *Sociobiology, the New Synthesis.* Belknap Press, Cambridge, Massachusetts
Wilson, S.C. 1982a. The development of social behaviour between siblings and non-siblings of the voles *Microtus ochrogaster* and *Microtus pennsylvanicus. Anim. Behav. 30* : 426-37
_____ 1982b. Parent-young contact in prairie and meadow voles. *J. Mamm. 63* : 300-5
Wittenberger, J.F. 1978. The evolution of mating systems in grouse. *Condor 80* : 126-37
_____ 1979a. The evolution of mating systems in birds and mammals. In P. Marler and J. Vandenbergh (eds.) *Handbook of Behavioral Neurobiology: Social Behavior and Communication.* Plenum Press, New York, pp. 271-349
_____ 1979b. A model for delayed reproduction in iteroparous animals. *Amer. Natur. 114* : 439-46
_____ 1981. *Animal Social Behavior.* Duxbury Press, Boston
_____ and Tilson, R.L. 1980. The evolution of monogamy : hypotheses and evidence. *Ann. Rev. Ecol. Syst. 11* : 197-232
Wolf, M., Schempp, W. and Vogel, W. 1979. *Ellobius lutescens* Th. (Rodentia, Microtinae); G-, R-, and replication banding patterns. Chromosome I polymorphism in the male and presumptive heterogamety in the female. *Cytogenet. Cell Genet. 23* : 117-23
Wolff, J.O. 1980. The role of patchiness in the population dynamics of snowshoe hares. *Ecol. Monogr. 50* : 111-30
_____ 1985a. Maternal aggression as a deterrent to infanticide in *Peromyscus leucopus* and *P. maniculatus. Anim. Behav. 33* : 117-23
_____ 1985b. Behavior. In R.H. Tamarin (ed.) *Biology of New World* Microtus. Special Publication No. 8 , American Society of Mammalogists, pp. 340-72
_____ and Lidicker, W.Z. 1980. Population ecology of the taiga vole, *Microtus xanthognathus*, in interior Alaska. *Can. J. Zool. 58* : 1800-12
_____ and _____ 1981. Communal winter nesting and food sharing in taiga voles. *Behav. Ecol. Sociobiol. 9* : 237-40
_____ and Lundy, K.I. 1985. Intra-familial dispersion patterns in white-footed mice, *Peromyscus leucopus. Behav. Ecol. Sociobiol. 17* : 381-4
Wolk, E. and Jozefczak, E. 1980. Activity of blood serum enzymes in seasonal generations of the bank vole. *Acta Theriol 25* : 377-89
Wright, S. 1978. *Evolution and the Genetics of Populations. Vol. 4. Variability Within and Among Natural Populations.* University of Chicago Press, Chicago
_____ 1982. Character change, speciation, and higher taxa. *Evolution 36* : 427-43
Wunder, B.A. 1985. Energetics and thermoregulation. In R.H. Tamarin (ed.) *Biology of New World* Microtus. Special Publication No. 8 , American Society of Mammalogists, pp. 812-44
Yamazaki. K., Beauchamp, G.K., Wysocki, C.J., Bard, J., Thomas, L. and Boyse, E.A. 1983. Recognition of H-2 types in relation to the blocking of pregnancy in mice. *Science 221* : 186-8
Yaskin, V.A. 1984. Seasonal changes in brain morphology in small mammals. In J.F. Merritt (ed.) *Winter Ecology of Small Mammals.* Special Publication No. 10, Carnegie Museum of Natural History, Pittsburgh, pp. 183-91
Yom-Tov, Y. 1975. Sychronization of breeding and intraspecific interference in the carrion crow. *Auk 92* : 778-85
Yu, O., Vergne, Y and Gounot, M. 1980. Modele d'interaction entre campagnols *Microtus arvalis* et prairie permanente *Rev. Ecol. (Terre Vie) 34* : 373-426

Zammuto, R.M. and Millar, J.S. 1985. Environmental predictability, variability, and *Spermophilus columbianus* life history over an elevational gradient. *Ecology 66* : 1784-94

Zejda, J. and Pelikán, J. 1984. Influence of radiation on the litter size of two microtine rodents. In J.F. Merritt (ed.) *Winter Ecology of Small Mammals*. Special Publication No. 10, Carnegie Museum of Natural History, Pittsburgh, pp. 235-41

Zucker, I., Johnston, P.G. and Frost, D. 1980. Comparative, physiological and biochronometric analyses of rodent seasonal reproductive cycles. *Prog. reprod. Biol. 5* : 102-33

Zullinger, E.M., Ricklefs, R.E., Redford, K.H. and Mace, G.M. 1984. Fitting sigmoidal equations to mammalian growth curves. *J. Mamm. 65* : 607-36

Zwickel, F.C. 1980. Surplus yearlings and the regulation of breeding density in blue grouse. *Can. J. Zool. 58* : 896-905

_____ 1983. Factors affecting the return of young blue grouse to breeding range. *Can. J. Zool. 61* : 1128-32

Index

Author's note: this book is designed to be read from cover to cover, and contains a number of interwoven themes which are awkward to index. The main themes of the book are reflected by the Chapter headings, and include the evolution of dispersal, territoriality and mating systems, and patterns of growth and reproduction. Food availability, habitat heterogeneity and population density are frequently implicated in this discussion. However, none of these important topics appear as major headings in this Index. The interested reader should tackle the lot.

α-selection 8, 109, 142, 146
Akodon 140
allometry 5-6, 31, 105, 176
Antechinus stuartii
    dispersal 156-7, 161-4, 167-8
    litter size 136
    stress 22-4
    winter aggregation 92, 167-8
Arvicola terrestris 30, 75, 77, 155
Ashmole hypothesis 134-5, 148

Bruce effect 55, 58-60, 91, 98, 116

canalisation 126
Chitty effect 108-11
Chitty Polymorphic Behaviour Hypothesis
    alternative formulation 186-7
    application to other data 109, 124
    behavioural evidence 15-17
    dispersal 154
    genetic evidence 10-15
    mathematical models 8-10, 146-7
    predictions 7-8
    problems 17-19
Clethrionomys
    gapperi 16, 30, 75, 82, 106, 123
    glareolus
        cost of reproduction 106
        diet 30
        dispersal 155, 160, 170
        genetics 14-15, 123
        growth 110-17 *passim*
        latitudinal gradients 4, 104
        litter size 131, 134
        mating system 75, 81-2, 89-90, 98-9
    rufocanus 30, 63, 75-7, 89, 110, 114, 160
    rutilus 30-1, 75, 82, 89, 114, 155
comparative method 75, 78, 87, 141

competition 40, 113
constraints 104
Coolidge effect 84
coprophagy 32
copulation 83-6, 99, 103, 168
costs of reproduction 77, 104-7 *see also* residual reproductive value, trade-offs
cyclicity
    latitudinal pattern 4-5, 14, 37, 39, 104, 108-11, 116, 120, 133-6, 144-6, 149, 170, 176
    mathematical basis 3, 5, 8, 35-6, 144-7, 176-8

developmental conversion 126-8
diapause 8, 187
Dicrostonyx
    torquatus 30, 139-42
    groenlandicus 120, 155
dispersal sink 152, 180, 182
DNA finger-printing 62
dominance 60-1, 70-1, 82-3, 92-6 *passim*

electrophoresis 12-15, 19-20, 62-3, 103, 122-3, 169-70
Ellobius lutescens 141
encounter experiments 15
endocyclical selection 122-3, 128
energetics 43, 91-2, 105-7, 122-3, 126, 166
environmental potential for polygyny 93-4, 136

fence effect 24, 180
frequency dependence 15, 58, 63, 95, 100, 166-7, 187

game theory, *see* frequency dependence

236

INDEX 237

General Adaptation Syndrome 20
genetic drift 12-14, 19-20 *see also*
    electrophoresis
grouse
    aggression 8, 10-14, 111, 186-7
    dispersal 159
    food 25, 36-7, 179
    mating systems 90, 95-6, 186-7
    time to reproduce 117
growth 43-6 *passim*, 107-11 *see also*
    puberty
gut morphology 32, 76, 132

habitat
    constant 48-9, 128, 156
    colonisation 49, 128, 155-6
    invasion 49, 128, 155-6
    traversible 49
    survival 49
habitat classification 42-7
habitat selection 40-2
heritability
    aggression 8
    dispersal 168-9
    good genes 71
    life history 107, 110, 122-9, 133
home range size 89, 187 *see also*
    territoriality

ideal free distribution 101
inbreeding 58, 66, 68, 71, 99, 114-15,
    132, 139, 141-2, 158-64
induced ovulation 53, 69, 86, 98
infanticide 53-8 *passim*, 67, 70-1, 86,
    90, 95, 174
intrademic group selection 141-2
islands 103-4, 122-3, 169

K-selection 8, 42-3, 142
kin recognition 19, 57, 65-9, 96, 114
kin selection 18, 65
Kin Selection Hypothesis 18-19
kleptogamy 84, 161

lactation 15, 52, 92, 98, 105-7, 121
laboratory studies, problems 101,
    113-14
Lemmus
    lemmus 30, 155
    sibiricus 30, 66
    trimucronatus 115
Lemniscus curtatus 75, 79, 87, 91-2

Lepus americanus see snowshoe hares
life-span 130-1
litter size 64, 70, 105, 111, 126, 131-7,
    148
local mate competition 141
local resource competition 138-9,
    166-7
lynx 177

management 38
Mastomys 3
mate choice 53, 61, 64, 71, 83, 98,
    158, 162-4
Meriones unguiculatus 139-40, 162
Microtus
    agrestis
        Bruce effect 55
        copulation 84
        diet 30-1, 110
        dispersal 103, 155, 160, 170
        genetics 15, 19
        growth 110, 116, 118-19
        habitat 47
        home range 89
        litter size 112, 134
        mating system 75, 98-9
        operational sex ratio 81
        photoperiod 90
    arvalis
        cost of reproduction 106
        diet 30
        dispersal 159
        growth 108
        litter size 131
        mating system 70, 75, 79, 87-92
            passim, 98-9
    breweri 30, 51, 103, 160, 169
    californicus
        competition 40
        copulation 85
        cost of reproduction 136
        diet 29-34 *passim*
        dispersal 160, 169
        genetics 19-20, 123, 133, 169
        growth 109, 111, 116, 121-3
        habitat 41, 46, 51
        incest avoidance 66
        litter size 131-3
        mating system 59-60, 75, 79,
            89-93 *passim*, 98-9
        operational sex ratio 81
        paternal care 86

predation 37
canicaudus 66-7, 84-5
montanus 3
    diet 179
    growth 119
    habitat 51
    mate choice 60-1
    mating system 75, 79-86 *passim*, 89-90, 98-9
ochrogaster
    Bruce effect 59-60
    copulation 84-6
    diet 30-1, 87-91, 132
    dispersal 155, 160, 169-70
    genetics 14, 16, 19, 169-70
    growth and puberty 53, 55, 114
    habitat 51, 87-91
    home range 89
    incest avoidance 66, 75
    litter size 32, 136
    mate choice 60
    mating system 64, 78-92 *passim*, 98-9
    paternal care 64, 91
oeconomus
    copulation 85
    diet 30
    dispersal 155
    habitat 51
    incest avoidance 66, 115
    mating system 75, 89
oregoni
    habitat 47, 51
    operational sex ratio 80, 82, 88-9
    sex determination 141
pennsylvanicus
    aggression 16
    communal breeding 92
    copulation 84-6
    cost of reproduction 106
    diet 30
    dispersal 24, 155, 160, 164-5, 169, 180, 182
    genetics 16, 62, 169
    growth and puberty 82, 109, 111, 138-9
    habitat 47, 51
    home range 89
    incest avoidance 66, 116
    longevity 131
    mating system 60, 75, 77-8, 81-6 *passim*, 98-9
    sex ratio 138-9
pinetorum
    copulation 85
    cost of reproduction 105-6
    dispersal 160
    growth 108, 113
    habitat 51
    incest avoidance 66
    litter size 136
    mating system 75, 79, 89-92, 98-9
    paternal care 64
richardsoni
    diet 30
    habitat 51
    home range 89
    mate-guarding 53
    mating system 75, 77
    sex ratio 139
townsendii
    aggression 10, 16
    dispersal 108, 159-60, 169
    genetics 169
    growth and puberty 108-10, 123
    habitat 47, 51
    mating system 74, 79, 89, 174
    operational sex ratio 82, 88
    predation 38
    simultaneous settlement 182
    stress 159-60
xanthognathus
    copulation 85
    diet 30
    dispersal 160, 164, 167
    habitat 47, 51, 128
    litter size 136
    longevity 131
    mating system 75
    winter aggregation 66, 164, 167
mid-winter aggregation *see* winter aggregation
mitochondrial DNA 62
monogamy 16, 52, 59, 64, 78-9, Chapter 4, 105, 113-14, 136, 158
    facultative 87, 136
    obligate 87, 136
MUS 64-5, 68, 71, 112-16, 122-3, 139
Myopus schisticolor 29-30, 139-42

Neotoma
    floridana 137
    stephensi 31

oestrus synchrony 63-4, 77, 93-4
olfaction 53, 68
operational sex ratio 80-3, 99, 167
optimum outcrossing distance 162-3
Oryzomys 3

parasites 38-9, 71, 154
paternal investment 59, 64-5, 84-7, 99, 136
paternity exclusion 61-3
Peromyscus
   eremicus 106
   floridanus 105-6
   leucopus 15, 51, 106, 160
   maniculatus 106, 138
   polionotus 19, 105-6, 121
persistence 143
Phenacomys longicaudus 29, 108, 136
phenotypic plasticity 126
phenotypic modulation 126-7
Phodopus sungorus 106
photoperiod 90, 99, 113, 119-20, 124-8 *passim*
Pitmys subterraneus 113
plant secondary compounds 5, 32-5
pleiotropy 123-8 *passim*
polyandry 78
polygyny 52
   male dominance 16, 78, Chapter 4
   mate defence 78
   resource defence 16, 78 Chapter 4
Polymorphic Behaviour Hypothesis *see* Chitty Polymorphic Behaviour Hypothesis
predation 35, 37-8, 92, 103, 134, 154, 164, 181
promiscuity 62
Pseudomys shortridgei 155-7
puberty 65, 72, 82, 98, 107-29, 138-9, 142-4

r-selection 8, 42-3, 109, 142
Rattus 3
recruitment, female control 82, 98
removal experiments 82, 88, 96, 102, 114, 124, 152
resource renewal 73-5, 97
risk aversion 127, 142-3, 148-9, 156
residual reproductive value 59, 105, 121-4, 135-6, 145-6

scent glands 15, 110-11
seasonal generations 108, 116-25, 138-9
seasonality
   demographic consequences 145, 147-9, 182-6
   dispersal 156, 172
   food 30-1, 36
   genetic consequences 68, 145
   habitat availability 40, 44-6, 50, 156
   litter size 131, 148
   reproduction 90, 94-5, 113-14, 176
   selection differences 58, 147-9, 182-6
   social consequences 94-5
senescence 64, 121-2
sex ratio 57, 81, 137-42, 166-7
sex determination 139-40
sexual selection 56, 70-1, 73 *see also* mate choice
Sigmodon hispidus 105-6
simultaneous settlement 182, 187
snowshoe hares
   aggression 10, 16
   dispersal 96
   food 25, 35-6
   habitat 36, 47
   mating systems 96
   population dynamics 117, 184-7
   predation 37, 177
   reproduction 117
sperm competition 61, 63, 67, 95
Social Fence Hypothesis 24-5
spring declines 24, 182
stress 5, 19-24, 39, 110, 124, 129, 132, 159-60
Synaptomys cooperi 160

t-alleles 68, 71
time to reproduce *see* puberty
trade-of 104
transfer 164-8
Trivers-Willared hypothesis 137-9

Whitten Effect 116
winter aggregation 66, 70, 91-3, 99, 136, 164-8, 174
winter breeding 90, 128, 184